Forces of Change

Events That Led to the Development of the Green Bay Fire Department

1836 - 1895

Washington Hook and Ladder Company No. 1 of the Green Bay Fire Department in 1881. This volunteer company formed in 1857 and became Truck No. 1 upon transition to the full-time, paid department in 1892. *(Green Bay Metro Fire Department archives)*

David Siegel

M&B Global Solutions, Inc.
Green Bay, Wisconsin (USA)

Copyright © 2016 David Siegel

First Edition

All rights reserved. With the exception of quoting brief passages for the purposes of review, no part of this publication may be reproduced without prior written permission from the Publisher or copyright holder. The information in this book is true and complete to the best of our knowledge.

Disclaimer
The views expressed in this work are solely those of the author and copyright holder and do not necessarily reflect the views of the publisher, and the publisher hereby disclaims any responsibility for them. In the event you use any of the information in this book for yourself, which is your constitutional right, the author, the copyright holder and the publisher assume no responsibility for your actions.

Published by M&B Global Solutions, Inc.
Green Bay, Wisconsin (USA)

Printed by Seaway Printing Company
Green Bay, Wisconsin (USA)

ISBN 10: 1-942731-20-5

ISBN 13: 978-1-942731-20-7

On the front cover:
Germania No.1 steamer fire engine in 1875. The Amoskeag Manufacturing Company of Manchester, New Hampshire, manufactured this steamer, formally named "Enterprise." *(Neville Public Museum of Brown County)*

On the back cover:
Green Bay Fire Station No. 3 about 1900: The City of Fort Howard built this station in 1873 for the volunteer fire department to replace one destroyed by fire. The hose cart is marked with the company number "3" and "G.B.F.D." This house served as GBFD Station 3 until 1937 when it was replaced by current Station 3 on Shawano Avenue. *(Green Bay Metro Fire Department archives)*

Contact the author at gbfd.history@gmail.com

Dedication

To Us

Especially those we've lost

Change: To make different in some particular, alter.
 —*Merriam Webster's Tenth Edition*

The most damaging phrase in the language is, "We've always done it this way!"
 —*Rear Admiral Grace M. Hopper, United States Navy*

Contents

Acknowledgements	viii
Foreword	x
Preface	xiii
1. Fire and Green Bay	3
2. Faltered Beginnings	9
3. Dawn of the Green Bay Fire Department	27
4. Transition to Steam Fire Engines	59
5. Consolidation and Progress	91
6. The Great Fire of 1880	115
7. Green Bay and Fort Howard Water Works Company	141
8. Birth of the Full-Time, Paid Green Bay Fire Department	167
9. Creation of the Fort Howard Fire Department	203

Contents

10. Gradual Changes
    The Fort Howard Fire Department — 229

11. Mutual Aid Leads to a Merger — 251

Epilogue — 283

Appendix I
Survivors of the Great Fire of 1880 — 287

Appendix II
Artifacts of the Early
Green Bay Fire Department — 293

Appendix III
Current Views of Former
Fire Station Sites — 305

Appendix IV
Green Bay Fire Department
Line-of-Duty Deaths — 315

Bibliography — 316

Endnotes — 321

Image Credits — 344

Index — 345

Acknowledgements

I've read a lot of history books and always passed over the acknowledgements. I didn't realize why the authors offered such generous thanks. After completing this project, I entirely understand why. A lot of people helped to make this book a success, and I really want to thank them.

Since original research is the foundation for this project, I'm profoundly grateful for the expertise, guidance, and knowledge provided by Mary Jane Herber, Brown County Library, Local History and Genealogy Department; Deb Anderson and her staff, University of Wisconsin-Green Bay, Area Research Center; and Louise Pfotenhauer, Neville Public Museum of Brown County. Without them, this would be a coloring book without lines.

As an amateur historian undertaking such a monumental task, I'm thankful for the professional mentoring from Dr. Adam Steuck, White Pillars Museum, De Pere Historical Society.

As an amateur writer, I'm deeply indebted to Mike Dauplaise, Bonne Groessl, and Amy Mrotek of M&B Global Solutions Inc. They made this work look and read as good as it is.

Along the same line, thanks to Jeremy Ness at Seaway Printing in Green Bay for patiently answering my questions.

Other thanks for research help go to Emily Turriff, Heritage Hill State Historical Park; Nancy Quirk and staff, Green Bay Water Utility; Chris Naumann and staff, On Broadway; Ian Griffiths, Berners Schober; Jim Hebel, Smet Construction; Julie Lamine, Meyer Theatre; Julie Tessmer and Heidi Yelk, Wisconsin State Law Library; Paul Archambault, Waterford Historical Museum and Cultural Center; Dr. John Arnold, NICOM, for research at the National Archives; as well as Jason Anderson and Michael Keane, Wisconsin Legislative Reference Bureau.

Generously providing and sharing images were Mike Hronek for the custom maps; Dennis Jacobs and his keen eye for photographs; Tom DeMeuse and members of the Algoma Fire & Rescue Association for access to the Smith pumper; Dick and Don Newman for their grandfather's Guardian No. 2 group portrait; family of Chief David Zuidmulder for the alarm boxes; Rick Fleury for the Franz Lenz certificate; family of Chief William Gleason for the use of his scrapbook; and Jason Flatt for maps of Green Bay.

I also want to acknowledge my Border Collie, Nith, who had many walks shortened or missed while I worked on this project.

Lastly, I want to thank all the firefighters, friends, and many others who expressed interest as this project progressed. Your enthusiasm bolstered my commitment.

David Siegel, Green Bay Metro Fire Department, Green Bay, Wisconsin

Foreword

It was with excitement that I finally held the finished manuscript of the book *Forces of Change: Events That Led to the Development of the Green Bay Fire Department*. For almost three years, David Siegel regularly came to use the research resources of De Pere Historical Society at White Pillars Museum, where I serve as the society's director. It was very clear that he undertook this project with enthusiasm and seriousness.

David always enters White Pillars Museum with both the energetic optimism of a young researcher and the grounded seriousness of a seasoned professional. I immediately noticed David organized his research well. (Frankly, many graduate students in history could take a lesson on the benefits of actually organizing research in folders with dates and subtopic dividers as opposed to piles of torn sheets of paper and Post-its that often decorate my own computer screen.) I quickly found it very interesting how David, who is a professional first responder, asked questions not about how past generations understood the threat of fire, or even historical content questions about aspects of Green Bay's history periphery to the fire department. Instead, he inquired a great deal about how professional historians undertake the research and writing process. The superior attention to this is the real strength of this book.

In *Forces of Change,* David follows a clear process of describing the evolution of the Green Bay Fire Department in the nineteenth century, and there is the thesis summarized rather simply, but completely. All too often, such local histories consist of a series of unrelated anecdotes that share a common anchor such as a geographic or constructed community, or an organization like a fraternal brotherhood or church congregation. These anecdotes might be interesting, and even well-researched. However, in most cases, you can remove entire chapters rather seamlessly and not notice. This is a telltale sign that what you are reading is not actually a history, but a collection of stories, and there is a difference. You would be hard pressed to remove a chapter from this book and maintain its effectiveness in explaining the changes to the Green Bay Fire Department. The reader needs to understand the growth of Green Bay, and to a lesser extent Fort Howard, to understand why the fire departments needed to change. However, one also must understand the benefits of transitioning from a volunteer to a professional fire department to grasp these changes, as well as the changes in firefighting technology. David covers all of these points and more not only with accuracy, but with balance to the overall thesis and purpose of the book. To remove one of these elements would degrade the clarity of the overall project. This is not a collection of stories; it is a history.

That is not to say there are no exciting accounts of fires in Green Bay, biographical information on Green Bay's *first* first responders, or detailed public debate between politicians and victims of destruction. These points are certainly present, yet all of these elements are there in the sense that they add to the larger understanding of the Green Bay Fire Department. They emerge in and out with the dexterity of a disciplined historian. Even the City of Green Bay itself is an ever-present, secondary character. By not bombarding the reader with endless data on the city, its buildings, and its people, we learn more about those very points than from many previous histories on nineteenth-century Green Bay.

When I write about others' research, I most often do so as a reviewer. Being asked to write the foreword to this book is an honor and a new experience for me. This is a book that explores new territory about the evolution of the Green Bay Fire Department, and perhaps more importantly, offers a unique approach to a local history that I hope to see replicated in future regional works.

Forces of Change leaves us with some good ideas of how firefighters in Green Bay adopted new technology and when they did; what led to professional firefighting department; and why preventative measures on building construction took place. The all-important *why* that runs through the entire book is key to creating new ideas that move us to ask new questions. Questions lead to discussion, and discussion leads to a clearer view of history. It is my hope that this book not only offers a view into the history of the Green Bay Fire Department, but that it also starts such a conversation.

Dr. Adam Stueck
Director, De Pere Historical Society

Preface

I am a firefighter/paramedic with the Green Bay Metro Fire Department (GBFD). This book is about the origins of that fire department. For my past, current, and future GBFD comrades, this story is about us, where we came from, and how our department came to be—long before any of us were born.

History can be intriguing, especially locations and objects, as these are tangible connections to the past. Plaques, markers, and signboards are irresistible to me. An old wall, bridge, or tower deserves a few minutes. A building takes even more. A Civil War battlefield, concentration camp, or museum requires hours. My travel companions know this very well.

This book tells the history of GBFD in terms of seminal, landmark events. For Green Bay residents, these events occurred in places we drive or walk by, where we work, socialize, or live. Part of the allure of history is to be at a site and know that long ago, something happened right there. This induces an effort to take our mind's eye back in time. To imagine the modern scene before us transposed over past events. To see firefighters crossing the Fox River on a ferry to help fight a fire on the other side. To watch firefighters haul a hand pumper fire engine through muddy streets, which we travel today as paved roads. To imagine a horse-drawn steamer fire engine pulling out of the fire station on South Washington. To stand on the very spots where significant (usually very large) fires shaped Green Bay and GBFD history.

These things happened in places we know well, not some distant, unfamiliar locale. Some of the fascination with history is to imagine this other world, left behind by time, within our own. Cast a backward glance.

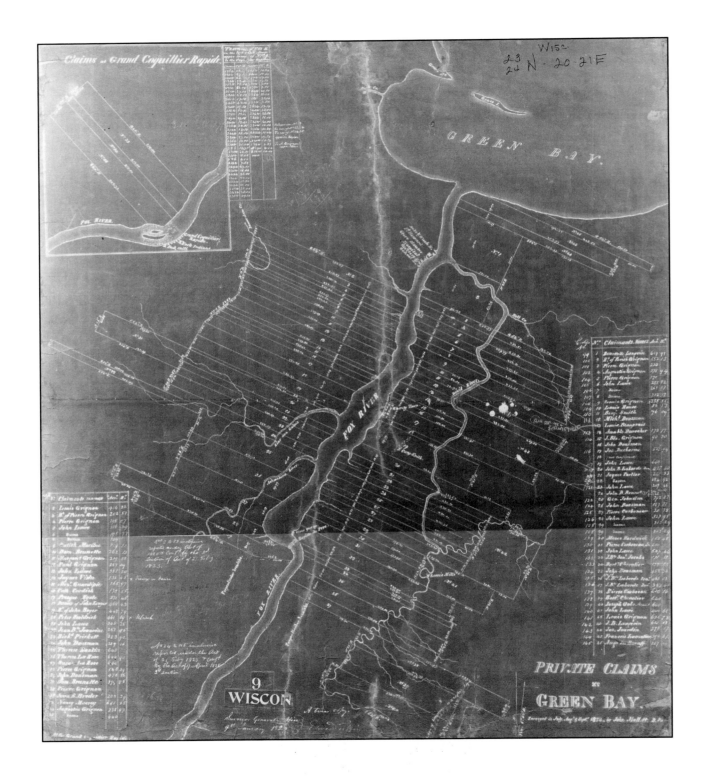

Chapter 1

Fire and Green Bay

Humans use fire as a tool to heat, cook, protect, illuminate, fabricate, build, and transport. Unique among tools, fire must be kept in tight check. Loss of control results in damage, destruction, harm, and potential death. An article in the October 27, 1859, *Green Bay Advocate* newspaper succinctly summarized, "Fire is a good servant but a hard master."

Firefighting is essentially a two-part, organized process—to both prevent loss of control and regain control (also known as suppression). As with most human endeavors, firefighting has changed over time. From simple community cooperation to sophisticated machines and dedicated, highly-trained professionals, methods of fighting fire have evolved through the centuries.

(Opposite) Private claims survey map of the Fox River area. This map was prepared in 1826, about ten years after US soldiers established the Fort Howard garrison, when there were no formal municipalities in the area. A few years after this map was created, the villages of Astor and Navarino were platted on the east side of the Fox River and then merged in 1838 to create the Borough of Green Bay. It was not until 1856 that the Borough of Fort Howard was incorporated on the west side.

Firefighting in Green Bay, Wisconsin, has undergone a relentless succession of transformations, most of which came about because of seminal or landmark events such as calamitous fires, mechanical innovations, municipal development, or economic realities. These pivotal events are the forces of change for the Green Bay Fire Department (GBFD).

Green Bay is a site that inherently attracts human communities. The land is fertile and resources available. Most importantly, the Fox River historically served as a connection between the Great Lakes and interior waterways, specifically the Mississippi River via the Wisconsin River. Whether with Native American tribes, French explorers, British settlers, or American pioneers, the Green Bay area has been an enticing place to live for centuries.

French explorers and fur traders were the first Europeans to establish western settlements here in the seventeenth and eighteenth centuries. Later, the British Empire governed the area, which was in turn usurped by Americans following the American Revolution, and in particular, after the War of 1812. Prior to becoming the northwestern frontier of the expanding United States of America, only a few hundred people lived in the area. Green Bay served as a trading post as well as a rest and supply spot for voyageurs preparing to venture even further. In 1816, the US Army built and occupied a garrison named Fort Howard on the west side of the Fox River, just south of the bay of Green Bay. The role of the solders at Fort Howard was to protect the important Fox-Wisconsin waterway and support the frontier communities. Settlers from the East soon followed, generating a sizeable influx of people into Green Bay.

The city of Green Bay began taking shape on the east side of the Fox River as the adjacent villages of Navarino and Astor. Both villages extended a few blocks inland from the river. Navarino, platted in 1829 by Daniel Whitney, was just south of modern East Walnut Street and extended north to the East River. Six years later, agents for John Jacob Astor laid out the village of Astor immediately adjacent to Navarino on the south. The villages merged in 1838 to form the Borough of Green Bay, which stretched just a few blocks east to west, and north-south from the East River to about modern Grignon Street.

Early inhabitants lived relatively isolated from each other in the heavy woods. As a result, fire destruction concerns were limited to individual dwellings. However, continual increases in population necessitated laying

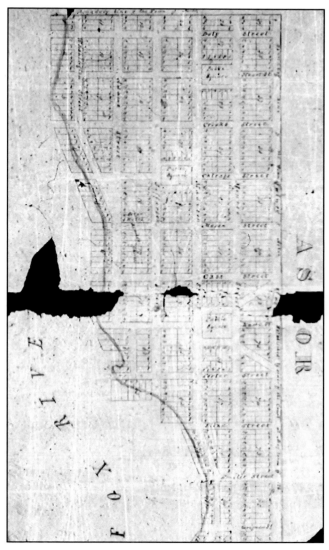

Plat map of the Village of Astor from 1835.
This map shows the streets of the current City of Green Bay from Walnut, south to Grignon and from the Fox River, east to Quincy.

Fire and Green Bay

more streets and constructing more dwellings, creating a more contiguous village. Municipal-wide firefighting consequently became a reasonable concern. As Green Bay grew from a dispersed wilderness settlement into a frontier town, it became possible for a fire at one location to spread to an adjacent property. Even more importantly, the larger population provided opportunities for residents to band together and successfully suppress uncontrolled fires. Within a few years, formal efforts began to create an organized force in Green Bay—a fire department.

Firefighting and the organization of the Green Bay Fire Department have not changed by chance or whim. A variety of forces influenced the department's evolution, with by far the most change occurring in the aftermath of significant fires. Hence, the lessons learned and the history of GBFD could be written symbolically with soot and ash.

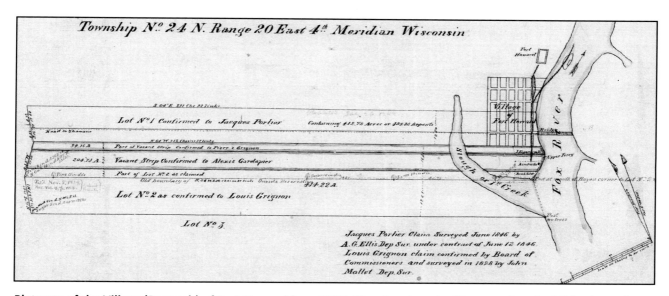

Plat map of the Village (Borough) of Fort Howard from 1868. The abandoned military garrison is visible above the street grids on the right side of the map, just west of the river. Also visible are the first railroad tracks into the area (1863) and the first bridge across the Fox River to Green Bay (1863).

Postcard of the Old Croc hand pumper fire engine. The US military placed this fire engine at the Fort Howard garrison prior to selling it to Green Bay in 1843. About six people on each side would move the long handles (known as brakes) in an up-down motion to power a pump located in the body of the machine between the two larger rear wheels. Water entered the pump through a supply hose attached to an inlet between the rear wheels (not visible). Water flowed from a single discharge at the top of the tower into a fire hose. People hauled this engine by hand using ropes stored in the reels between the smaller front wheels. Others steered using the upright piece, attached to the front axle, after it was lowered.

Chapter 2

Faltered Beginnings

1836 - 1853

On January 5, 1836, Daniel Whitney chaired a meeting of the citizens of Navarino, which two years later would become the North Ward of Green Bay. Several resolutions were passed "in order to better guard against fire" in the recently created village. Although non-binding, these resolutions encouraged each house to keep readily available two fire buckets to throw water and at least one ladder to reach burning roofs. Additionally, a committee was formed to procure money for dedicated firefighting equipment, such as fire hooks, ladders, and communal buckets, as well as to estimate the cost of a fire engine.[1] It is unknown what, if any, equipment was actually procured, but it is certain a fire engine was not obtained at this time. Regardless of success, these were the first efforts to establish fire suppression in the frontier community of Green Bay.

Reward announcement from the February 19, 1834, *Green Bay Intelligencer*. This attempted arson attack predated creation of the Borough of Green Bay and an organized fire department.

> **Fire! Fire!! Fire Buckets!!!**
> THE subscriber will make to order Fire Buckets, and all persons wishing them made; can be supplied upon application to   E. W. BANCROFT.
> Green Bay, June 25, 1839.

Advertisement for fire buckets from the June 25, 1839, *Wisconsin Democrat*, a Green Bay newspaper. Citizens used buckets to collect water from rivers or wells and then throw the water on the fire. These were the only firefighting tools in the frontier town and an ordinance required each household to have buckets available.

As part of the process to become a legally recognized community, the Wisconsin territorial authorities approved a charter in 1838 for the new Borough of Green Bay. The charter listed the numerous powers and responsibilities of the borough's president and trustees. Included were two brief provisions giving municipal officials the legal basis to implement systematic firefighting in Green Bay. The charter gave authorities the power "to establish a fire department, and provide for the extinguishment of fire; to regulate the storage of gunpowder, and other combustible materials."[2]

This simple act recognized both aspects of firefighting—suppression and prevention. Shortly afterward, the first detailed fire ordinance required every household to have at least two buckets readily available "in case of accident by fire and for no other purpose." Failure to comply resulted in a $5 fine. Within a year, officials modified this ordinance to require that buckets be numbered and "painted with the owners' name."[3]

Given the small population and absence of an organized fire department, the frontier community of Green Bay implemented a citizen-based bucket brigade system. Very simply, lines of people would extend from the water source (a well or natural body of water) to the burning

> **AN ORDINANCE**
> *In relation to Fire Buckets.*
>
> At a subsequent meeting of the Board of Trustees held at the store of Daniel Butler, on Saturday the 25th day of August, 1840, it was
>
> "*Resolved,* That the Fire Buckets required by the "Ordinance to prevent injuries by Fires," may be made of wood, provided the same be numbered and painted with the owners name."
>
> A true copy from the original records.
> Attest; G. McWilliams, Clerk T. G. B.

Ordinance requiring labeling of fire buckets from the January 1, 1842, *Green Bay Republican*. Without an organized fire company, members of the community combined resources and cooperated to fight fire. Apparently there were so many buckets in Green Bay that after the fires were extinguished, this system was needed to return fire buckets to the rightful owners.

building. Full buckets would be passed from person to person to the fire, where they would be dumped to douse the flames. The empty buckets would be returned to the water source in the same manner. This was a manpower-intense operation, making it imperative that many neighbors assemble and work together to fight the fire. Because a large number of buckets were needed, the ordinance explicitly mandated that all buckets be ready and accessible.

In addition to fire suppression, the first ordinance also addressed fire prevention. A committee of two was appointed "to adopt and enforce all necessary means for prevention of fire." The committee had the authority to take down any building it deemed to pose an excessive fire hazard. Building owners and occupants were vaguely required to "render safe from accident by fire, any fireplace, chimney, stove or stovepipe."[4]

It didn't take long for significant fires to strike nascent Green Bay. In late 1840, five large buildings and two small structures were destroyed in an inferno on the riverbank along North Washington Street.[5] One witness later wrote this was "the first great fire which this place has experienced," while another description stated the flames "destroyed a large portion of the business district of Green Bay" and "consumed much valuable property."[6]

Undoubtedly, the tremendous scale of the fire overwhelmed the efforts of the citizen bucket brigade.

An even more massive fire broke out on December 24, 1841, in the very same part of the mercantile area. First tearing through a three-story wooden building, the blaze soon spread relentlessly, eventually destroying an entire block of buildings that included several stores, the *Phoenix* newspaper office, papers, and maps of the district surveyor, as well as all the books, records and papers of the Brown County Sheriff.[7] Twice in the span of about a year, major fires had devastated young Green Bay.

A noteworthy event occurred during the December 1841 fire. As reported by Captain Ephraim Shaler, an army officer and caretaker for the then-inactive Fort Howard US Army garrison, a group of men from Green

Bay crossed the ice-covered Fox River to fetch the hand pumper fire engine kept at the fort. Shaler let them take the fire engine across the ice to Green Bay "without hesitation." He accompanied the fire engine and acknowledged they "succeeded in saving several valuable buildings."[8] For Green Bay, this was the first act of mutual aid—sharing firefighting resources between communities, a practice that thrives today.

The citizens saw firsthand the benefits of a fire engine versus a bucket brigade. The hand pumper fire engine was essentially a lengthy, wagon-like box mounted on four wheels. People would push and pull a pair of long handles, known as brakes, in an up-down, seesaw motion. Levers attached to the brakes propelled the water pump mechanism. Water was drawn from a cistern, well, or natural body through a short supply hose into the fire engine mechanism and propelled out a single discharge to a fire hose.

About twelve people simultaneously operated the brakes on the hand pumper brought over from the Fort

Old Maid of the New York Fire Department, a sister-engine to Old Croc. These were two of only three fire engines made by Harry Ludlum in the 1820s in New York.[9]

Faltered Beginnings

PUBLIC MEETING.

At a meeting of the citizens of Green Bay convened at the Navarino Hotel, on Friday the 24th December, 1841, for the purpose of taking measures for the protection of property by fire in said town, DANIEL WHITNEY, was called to the Chair, and F. GILBERT, appointed Secretary. The object of the meeting having been stated by the chairman, it was on motion

Resolved, That the formation of a company of firemen is essentially necessary for the protection of our property, and the safety of the town.

Resolved, That a committee of three persons be appointed by the chair, to call upon Capt. Shaler, and make arrangements with him, as the agent of the U. States, in relation to the Engine at Fort Howard, either for the purchase, or temporary use thereof.

Resolved, That a committee of three persons be appointed to receive subscriptions for the erection of an engine house, to be located in a central situation, in the Borough of Green Bay, and that said committee be authorized and requested to circulate a paper and obtain signatures of persons willing to become members of a fire company.

Resolved, That the Engine and apparatus shall be and remain in the care, and under the control of Col. C. Tullar, and Maj. W. C. Disbrow, until the formation of said company.

The resolution having been seperately considered, were unanimously adopted by the meeting.

Whereupon the chairman appointed H. S. Baird, W. Mitchell and W. H. Bruce, the committee to call upon Capt. Shaler, in regard to the leasing or purchasing the Engine.

Messrs. N. Goodell, C. Tullar, and A. J. Irwin, committee to circulate a subscription for the erection of an Engine house.

On motion it was further

Resolved, That a committee of three persons be appointed Fire Wardens.

Resolved, That the President and Trustees of the Town be requested to pass an Ordinance, to carry into effect the above resolution.

Resolved, That H. S. Baird, C. Tullar, and O. P. Knapp, be requested to act as Fire Wardens, for the North Ward, and A. J. Irwin, N. Goodell, and J. F. Lessey, for the South Ward.

Resolved, That the proceedings of this meeting be signed by the Chairman and Secretary, and published in the Green Bay Republican.

DAN'L WHITNEY, Ch'n.
F. GILBERT, Secretary.

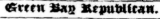

Saturday, January 1, 1842.

Howard military post. The device normally would operate at sixty strokes per minute, but could increase briefly to about double that speed. At the normal rate, a person would last about ten minutes and then need to rest.[10] Thus, a fairly large number of people were required to successfully operate the hand pumper. While some operated the brakes, others revitalized and waited for their turn. A lucky few held the nozzle and directed the water stream, although they, too, likely took turns operating the brakes. The effectiveness of the hand pumper fire engine had now been singularly demonstrated to the citizens of Green Bay, as was the need for a systematic firefighting system.

Whitney, who owned the largest building destroyed by the blaze, chaired a town meeting the day after the fire. The purpose of this meeting was solely for "taking measures for the protection of property by fire." Attendees passed several resolutions, including the formation of a company of firefighters, establishing a committee to approach Captain Shaler at the fort to make arrangements for purchasing, or at the very least, gaining temporary use of the hand pumper, raising funds to build an engine house on the east side of the Fox River, and appointing a committee of three fire wardens with a supporting ordinance that defined their duties.[11] For the first time, efforts were being made to coordinate firefighting in Green Bay.

Individuals and committees took action on these resolutions immediately. Within a few days of the town meeting, a committee of three proposed in writing that the military either sell or loan the fire engine to Green Bay. Since the hand pumper required a considerable number of people to operate, the proposition noted that because the military post was essentially abandoned, there were "no persons residing in the Fort, and the engine is entirely useless when inside the garrison." However, the citizens of Green Bay would "afford every assistance in our power in case of fire, at the Fort or in its vicinity."[12]

Shaler promptly informed his superiors he had already consented to leave the fire engine in Green Bay and even recommended selling it to the borough. The sale was readily approved by the Quartermaster in Washington, D.C.[13] However, Shaler reported in September 1842 that the fire engine had not been sold because Green Bay borough authorities had not accepted the offered sale price of $350.[14] Consequently, superiors instructed Shaler to send the fire engine to Fort Winnebago near modern-day Portage, Wisconsin.[15]

(Opposite) Results from a public meeting following the disastrous fire on North Washington from the January 1, 1842, *Green Bay Republican*. Green Bay founder Daniel Whitey chaired this meeting, not surprising given he lost several buildings in the December 24, 1841, fire. They held this meeting "for the purpose of taking measures for protection of property by fire." The outcomes of this meeting included the first organized firefighting force in Green Bay.

Faltered Beginnings

Fortunately for Green Bay, the order came late in the year. Shaler could not send the fire engine to Fort Winnebago, as there were no more boats that season, nor could it travel overland due to bad roads.[16] This delay gave Green Bay another opportunity.

On November 23, 1842, a month after Shaler wrote his superiors about the impasse, Whitney sent a letter on behalf of the borough trustees directly to the Secretary of War offering $350 for the fire engine, the price proposed by Shaler earlier that year.[17] Unfortunately, no records exist directly confirming this sale, although a February 1847 newspaper account clearly states a fire engine was used at a fire in Green Bay.[18] Thus, it is reasonable to conclude that sometime in 1843, sale by the military was consummated and the Borough of Green Bay formally possessed its first fire engine.*

Another resolution from the December 1841 meeting concerned the construction of an engine house. The borough raised money through subscription (promises to pay) to cover the cost of construction. While no information describing the building itself or date of construction is available, the first reference to an engine house in Green

---

*This first fire engine still exists – in great condition – and is on display at the Neville Public Museum in Green Bay. Known as "Old Croc," it was one of only three manufactured by Harry Ludlum in New York City in the 1820s.[21]

Bay comes from an 1847 newspaper article.[19] A later recollection states that by 1844, the Old Croc hand pumper responded from the engine house at the southeast corner of Walnut and Adams streets.[20] This first Green Bay fire station likely was a simple wooden structure, the predominant type of building then, and most likely constructed in 1842-43.

In addition to the efforts made toward fire suppression, Green Bay also took steps to prevent uncontrolled fire. The borough created fire warden positions at a December 24, 1841, meeting, and the very next day passed an ordinance specifically describing and formalizing their authorities and duties. Six citizens received appointments and were required to inspect each house, store, and other buildings once every two months. During these visits, the fire wardens examined chimneys, fireplaces, stoves, and stove pipes for unsafe conditions, and then reported their findings to the occupants. Refusal to allow fire wardens access or failure to correct dangerous conditions within forty-eight hours resulted in a $5 fine per refusal and another $5 fine per day for neglected repairs. Merchants were restricted to twenty-five pounds of gunpowder, which had to be kept in tin canisters.

Along with fire prevention roles, the wardens held authority to direct bucket brigades and fire engine operation at fire scenes. These fire wardens became the first firefighting officials in Green Bay. They wore caps with distinct insignias and held four-foot staffs at fire scenes to affirm formal authority and ensure recognition. Failure to comply with a warden's directions resulted in a $2 fine.[22] It is unknown how often these fines were imposed, but what is certain is a fire department authority had been established.

The December 1841 meeting concluded with a resolution to consider forming a fire company. Some evidence corroborates that a company was indeed formed. A committee soon circulated a petition to "become members of a fire company."[23] A few months later, as part of "An Act to Change the Corporate Limits and Powers of the Town of Green Bay," the municipality was authorized to form fire companies.[24] Most convincing of all was the letter from Whitney offering to buy "Old Croc," stating, "There is a very good fire company formed here [Green Bay]," solidly confirming there was an organized firefighting force by late 1842.[25]

By the beginning of 1843, the Borough of Green Bay had a hand pumper fire engine, an engine house, a fire

Election of officers of the newly organized Green Bay Fire Company No. 1 from the February 18, 1847, *Green Bay Advocate*. This company formed because the group organized in 1842 had ceased to function. This was the first named fire company in Green Bay.

Fire company meeting announcement from the February 25, 1847, *Green Bay Advocate*. These events included both training and social activities.

company, and six fire wardens with authority to administer both prevention and suppression activities. Managed firefighting in Green Bay had begun.

Unfortunately, it appears the efforts to maintain a fire company in Green Bay came undone. In a February 11, 1847, editorial, the *Green Bay Advocate* ran a mock story reporting a fictional fire consuming much of Green Bay. The paper lamented Green Bay had "a good engine, but no organized company, and it is not in a condition for use."

Apparently, the intentions of early 1842 had not come to fruition. The initial fire company no longer existed and "Old Croc" was out of operation, most likely due to lack of maintenance. The hand pumper's internal leather parts were particularly susceptible to failure as a consequence of a lack of attention and use.

Although sarcastic in tone, the true purpose of this editorial was to encourage attendance that night at a meeting in the engine house. Touting success afterward, the paper reported, "Green Bay Fire Company No. 1" had been formed at the meeting and officers elected.[26] The new fire company received formal sanction a year later when the Green Bay Board of Trustees passed an ordinance "To Organize Fire Company No. 1 in Green Bay."

This document recognized nine men as the initial members of the community's first named fire company. The ordinance also stated the company shall consist of "twenty-four to forty able-bodied men, between the ages of eighteen and fifty years." Membership would be voluntary and the company governed internally.[27] This second effort to start a fire department in Green Bay was more formal.

However, just like five years earlier, this fire company did not prosper. Notices announcing Fire Company No. 1 meetings appeared in the paper during the first year, but those notices no longer appeared after April

Ordinance officially recognizing Fire Company No. 1 from the April 13, 1848, *Green Bay Advocate*. This was the first official recognition of a fire department in Green Bay. Unfortunately, this fire company ceased operations within two years, leaving Green Bay without organized fire protection.

1848.[28] More telling was a September 1849 newspaper account reporting the destruction by fire of Elisha Morrow's house, which was under construction. The fire, caused by a worker using a furnace to prepare eave troughs, was fought "by workmen and citizens."[29] Notably absent was any mention of involvement by Fire Company No. 1.

Similarly, in March 1850, a major fire destroyed several downtown businesses and again, reports did not mention efforts by the fire company. Not only was there no mention of the any firefighter force, the paper lamented, "The engine belonging to the village being found useless."[30]

This time, soldiers and officers from the recently re-garrisoned Fort Howard came across the river with their own fire engine to provide aid.[31] Army personnel had reoccupied Fort Howard following the Mexican-American War. The army provided the fort with a replacement fire engine in 1850, because the previous hand pumper (Old Croc) had been sold to Green Bay seven years earlier.[32]

As before, this second effort to establish a fire company in Green Bay failed within a few years. Similar to 1841, the major fire in 1850 prompted a citizens' call to action. A public meeting was held on March 21, 1850, to

Fire report headline from the March 21, 1850, extra edition of the *Green Bay Advocate*. Green Bay did not have an organized fire company and the Old Croc hand pumper malfunctioned due to lack of maintenance.

Faltered Beginnings

(Opposite) The James Smith hand pumper in Algoma, Wisconsin, in 2015. Green Bay sold it to the village of Ahnapee (today Algoma) in 1875. The Smith had a much greater pump capacity than Old Croc. The longer handles (brakes) accommodated about twenty people on each side. There are two, 2½-inch discharges mounted on the front of the box above the smaller front wheels. The two rigid, black supply hoses along the sides above the wheels were attached to the rear inlet and lowered into the water source, usually a river. As with the Old Croc, firefighters hauled it by hand using ropes deployed from the reels and steered with the bar attached to the front axle.

first thank the numerous soldiers and officers for their help, and more importantly, to promote the opinion "that a regular and efficient organization of a fire department in this Town is essential."[33] The newspaper reported the borough trustees were encouraged to take several additional measures, specifically levying a tax to defray the expenses of a proper fire department.

This strongly suggests the failure to maintain a fire company after the 1841 resolutions and the Green Bay Fire Company No. 1 in 1847 was due to a lack of financial support. Additionally revealing was another proposal by the newspaper suggesting "two good engines are actually necessary" so as to "keep up an efficient department – which cannot be kept without that spirit which is generated by competition."[34]

It appears the absence of competition may have contributed to the failures as much as the lack of financial support. Much like friendly rivals, fire companies compete with each other to be first on the scene, first to spray water, etc. In spite of the efforts by the newspaper, a February 1851 article described two more fires. But again, there was no mention of fire company participation.[35]

In spite of two attempted starts, several substan-

Faltered Beginnings

> A NEW FIRE ENGINE, of the "piano" style, has landed from the *Michigan*, yesterday morning, and is a handsome looking machine. With a bran new company, formed expressly for it, under the foremanship of Dr. BELL, there won't be much use in a fire trying to go far around this burgh—we expect.

Arrival of the new fire engine celebrated in the May 15, 1851, *Green Bay Advocate*. The Green Bay Common Council purchased the hand pumper from James Smith of New York, then immediately gave it to the recently formed Alert Fire Company.

tial fires, and mounting public pressure, Green Bay still remained without a fire department.

Eventually, the Common Council took the initiative to promote a fire department. Likely prompted by the devastating mercantile district fire in March 1850 and a push to promote – again – the creation of a fire company, the Borough of Green Bay obtained a new "piano style" hand pumper fire engine in May 1851 manufactured by James Smith of New York.*[36] In fact, the newspaper stated a new fire company was formed expressly in anticipation of the arrival of the hand pumper.[37] Taking the name Alert Fire Company, the new unit featured at least thirty members.

Green Bay sold the Smith hand pumper in 1875 to the Village of Ahnapee, Wisconsin, about thirty miles to the east on Lake Michigan.[39] Ahnapee later would become the city of Algoma. The Smith hand pumper remains a prized possession of the Algoma Fire & Rescue Association, which proudly brings it out for parades.

They marched in an 1851 fall parade showcasing "their new Engine and Hose Carriage and new uniform."[38]

Not much is known about this group, but in the spring of 1852, the Alert Fire Company was praised for "the skill and energy displayed" and that they "more than paid for their engine in the saving of property" at two separate fires.[40] Unquestionably, the Alert Fire Company was successful.

As happened after the enthusiastic initiatives of 1841 and 1847, the Alert Fire Company soon faded away. An August 1852 newspaper notice announced a regular meeting, but there were none thereafter.[41] Less than a year later, an April 1853 newspaper article reported a fire at the Washington House hotel without mention of Alert Fire Company's involvement.[42] Similarly, in July of that year, a blacksmith shop on Washington Street was destroyed by fire. Again, there was no fire company response noted.[43]

Organized firefighting in Green Bay had failed to flourish for the third time in two years. However, events would soon change that for good.

Meeting notice in the January 29, 1852, *Green Bay Advocate*. There is some uncertainty regarding the official fire company name. The newspapers inconsistently reported the fire company name: sometimes No. 1, other times No. 2, and sometimes without any company number. A November 1853 fire destroyed all official municipal records referring to Alert Fire Company. The hand pumper icon is very similar to the Old Croc, in particular the discharge on top of the rear housing. However, Alert Fire Company used the Smith hand pumper.

Meeting announcement in the August 24, 1852, *Green Bay Spectator*. Within a year of this meeting, Alert Fire Company ceased to function becoming the third organized fire company to fail.

Faltered Beginnings

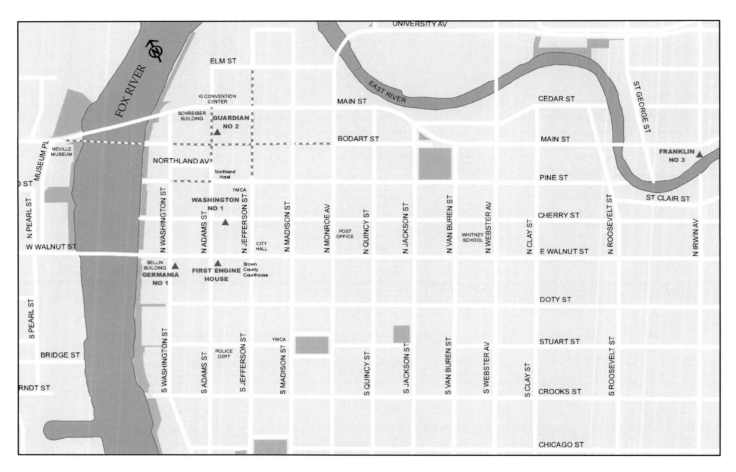

GBFD fire station locations (1842-1863) within modern Green Bay. Fire stations discussed in this chapter are indicated by the triangles and labeled with the company names. Dashed lines indicate streets that no longer exist as of 2016.

Chapter 3

Dawn of the Green Bay Fire Department
1853 - 1863

By 1853, Green Bay had a population of over 2,000 and was evolving from a frontier wilderness town to a proper city.[1] There had been three failed attempts to create a fire department since 1841, leaving Green Bay still without systematic fire protection when a series of three serious fires struck its commercial center. These events provided the forces of change necessary to finally establish a permanent Green Bay Fire Department (GBFD).

The prosperous business district, centered on Washington Street, was comprised of wooden buildings that stood in close proximity to each other, if not directly abutting. These congested and readily flammable structures provided ideal conditions for a disaster.

Seal of the Borough of Green Bay. This is attached to the Common Council meeting minutes from December 28, 1853.

> *"It is almost a positive certainty that two well organized Fire Companies, properly manned and worked, would have saved a large amount of property hopelessly lost."*
>
> - *Green Bay Advocate* editorial

In the early morning of November 1, 1853, a fire broke out in an unoccupied building between the Fox River and North Washington, just north of Pine Street. A "strong gale" from the west quickly fanned the flames, sending burning debris onto nearby buildings. These also caught fire, and in short order, so did all the buildings on both sides of Washington, spreading out for a block north of Pine. The inferno then spread eastward, burning most of the buildings on the block between Washington and Adams streets north of Pine. To the west, the fire burned every structure along the river, destroying the same one-block stretch of North Washington. Overall, thirty buildings, two freight boats, and about 20,000 shingles were destroyed. One account stated, "The best portion of the village is in ashes—some of largest and finest buildings are destroyed."[2]

The burned buildings were valued at nearly $100,000 and included a hotel, three warehouses full of stock, the office of the *Green Bay Advocate* – at that time the only newspaper in Green Bay – many stores and offices, as well as the municipal records dating to creation of the borough in 1838.[3] The destruction of Green Bay's commercial heart was serious enough, but even worse was the loss of supplies stockpiled for the coming winter.[4] In addition, one person was missing. His body eventually was found in the Fox River five months later.[5]

Newspaper accounts provide great detail on the course of the fire. The "violent wind . . . sent a constantly increasing current of heat, together with a thick shower of sparks, mingled with large pieces of blazing wood" to create a firestorm that spread relentlessly. It appears the fire ran its course and ended only when there was nothing left to burn. Essentially, the streets acted as fire breaks, depriving the flames of fuel. The only mention of firefighting was a note that one store was saved "with the aid of the engine and by the efforts of citizens." Another commentary plainly states Green Bay had a fire engine, but "no company to work it."[6] Criticism of the lack of a fire company was well-justified.

However, even this disaster did not immediately stimulate renewed efforts to form a fire department. The *Green Bay Advocate*, destroyed by the fire, returned to operation six weeks later and presented an editorial with an adamant message encouraging the formation of two fire companies. "It is almost a positive certainty that two well organized Fire Companies, properly manned and worked, would have saved a large amount of property hopelessly lost."[7] In addition, the paper directly implored citizens and

Headline from the December 15, 1853, *Green Bay Advocate.* The November 1, 1853, blaze destroyed the offices of this newspaper, which returned to print with this edition six weeks later.

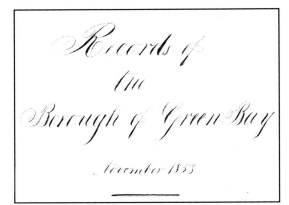

Inside cover of the Borough of Green Bay Common Council meeting minutes. The first recorded meeting is from November 19, 1853. All previous municipal records were destroyed in the fire eighteen days earlier.

borough authorities to take action. However, the Common Council meetings were devoid of any effort to promote or form fire companies. Some minor actions were taken to maintain the fire engine and hose, but nothing to actually bring a fire department into existence.[8] Astonishingly, in spite of the catastrophic November 1853 fire, Green Bay remained without a fire department.

Almost exactly a year after the previous conflagration, two more fires would strike Green Bay businesses at the end of 1854. These incidents, occurring so soon after the 1853 fire, ultimately spawned the forces of change that, for the fourth time, would create a structured fire company. This time, however, the formation of a fire department would be permanent.

On November 28, 1854, four stores on the west side of North Washington at Pine burned, as did a docked boat. Although not nearly as severe, this was the same location of origin as the massive conflagration that had occurred a little more than one year earlier. In describing efforts to combat the flames, the *Green Bay Advocate* reported the fire was too hot to use buckets. More telling, "There seems to have been great difficulty and delay in getting the engine into operation." Again, the fire was essentially allowed to burn unimpeded by any substantial suppression efforts.

Fortunately, the winds were light and a repeat of the devastating firestorm that happened one year before did not occur.[9]

Exactly one week later, the prominent Washington House hotel, located on North Washington Street at what was then Main Street, burned to the ground. Citizens, rather than a coordinated fire company, brought the engine to the fire. Even then, the hotel was lost to the flames. Most significantly, a man visiting from Little Chute heard of a young girl trapped upstairs and ran into the burning building. He suffered burns to his hands, and while both nearly suffocated, they survived.[10] Green Bay had experienced three major fires in just over a year, all without any organized firefighting response.

Having just avoided another conflagration in the commercial district, a destroyed hotel and a young girl's near death – all occurring without a fire department in place – the citizens reached a new level of intolerance and took determined action. At the December 2, 1854, Common Council meeting, numerous citizens presented a petition asking for the appointment of "a Watch as a more efficient protection against the danger of fire." The mayor received authorization the next month to appoint a watchman, supported by the necessary ordinance.[11]

"There seems to have been great difficulty and delay in getting the engine into operation."

- Newspaper account of the November 28, 1854, fire

Dawn of the Green Bay Fire Department

Most importantly, the Common Council went even further. At the December 7, 1854, meeting, the city marshal was authorized "to deliver the Fire Engine to any Fire Company that may organize before the next regular meeting of the council."[12] The authorities were blatantly soliciting for a fire company, and a positive response was immediate.

A group of German immigrants held a meeting on December 14, 1854, during which they prepared and signed a constitution forming Germania Fire Company No. 1.[13] With sixty members including elected officers, Germania No. 1 immediately was given the five-year-old Smith hand pumper.[14] Moreover, the company recommended, and the Common Council approved, F. A. Lathrop as chief fire warden in charge of the fire department.[15] The fire company gave itself the nickname "Rough and Ready," an obvious homage to Zachary Taylor. "Old Rough and Ready" had served as commander of the Fort Howard garrison and later was elected twelfth president of the United States. He passed away on July 9, 1850, shortly after taking office.[16] The immigrant firefighters of Germania No. 1 likely were inspired by a patriotic commitment to their new community. For the fourth time, Green Bay had an organized fire company, and this time Germania No. 1 would persist.

Announcement that Germania Fire Company No. 1 has organized from the December 21, 1854, *Green Bay Advocate*. Between 1841 and 1851, three Green Bay fire companies had formed, but each ceased operations within a couple of years. Following three serious fires in Green Bay over the previous year, German immigrants established Germania No. 1 on December 18, 1854. The lineage of this fire company can be traced directly to Green Bay Fire Department Engine Company No. 1, which operated until December 2012, exactly 158 years later.

In addition to using the Smith hand pumper fire engine, Germania No. 1 occupied the relatively new engine house on the east side of South Washington, just south of Walnut Street. Due to the loss of many municipal records in the November 1853 fire, the build year of this new engine house cannot be precisely established. However, some clues exist.

First, a June 1854 newspaper article refers to the "Old Engine House" (built around 1842) behind the Town House on Adams Street.[17] If the first engine house on Walnut and Adams was considered old in 1854, then there must have been a new one. The next clue comes in March 1856, when the city paid a long-overdue bill for a deed claiming the northern portion of a lot on South Washington—the site of the new engine house.[18] Even more telling are the actions of Burley Follett, who had presented the bill. In 1852, he owed taxes only on the southern portion of this lot.[19] The engine house was on the northern portion. From this evidence, it can be inferred that the second engine house was built in 1852 or earlier. It was this engine house that Germania No. 1 occupied with the Smith hand pumper in late 1854.

After Germania No. 1 was formally recognized and became active, community acknowledgment and support

Early Green Bay Fire Stations

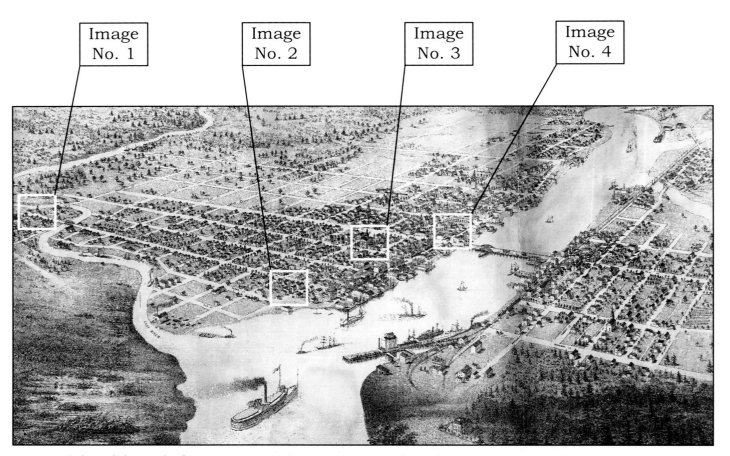

Birdseye lithograph of Green Bay in 1867. The highlighted boxes are explained in detail on the following page.

34   Chapter 3

Early Green Bay Fire Stations

1 - Franklin No. 3 Engine House. This fire house was built in 1860 for the newly created fire company on the south side of Main Street at the intersection with modern Irwin Avenue. The East River is behind.

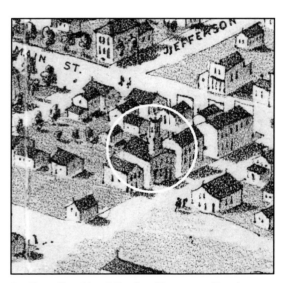

2 - Guardian No. 2 Engine House on North Adams. The street labeled "Main St." and this section of North Adams were eliminated by redevelopment in the 1970s. Modern Main Street shifted one block to the north, shown on the lower left of this image. This engine house was built in 1859. The site today is within the Schreiber Foods campus, which opened in 2014.

3 - Washington No. 1 House. This station was built in 1860 after it became apparent the North Adams station could not accommodate two companies. The "Hooks" house was on Cherry Street between North Adams and North Jefferson. The large building next door was the Brown County Court House.

4 - Engine house on South Washington. Germania No. 1 operated from this engine house (built about 1852) using the Smith hand pumper. The upstairs served as meeting and social rooms. This site is currently the Backstage at the Meyer, which opened in 2015. The east end of the Walnut Street Bridge is in the bottom right.

Dawn of the Green Bay Fire Department

> *"It is now almost impossible to get property insured in a reliable company."*
>
> - *Green Bay Advocate* editorial, in arguing for a second fire engine

immediately followed. The *Green Bay Advocate* promoted a "first annual Ball" as a fundraiser for the fire company.[20] Others complimented the fire company for being structured and spirited, and the paper expressed "hope at last to have a well-organized Fire Department."[21] Along with praise, the Common Council provided financial backing. For example, during the first winter Germania No. 1 operated, the Common Council paid to heat the engine house in order to keep the fire engine from freezing.[22] Also approved was the purchase of 200 feet of hose and numerous repairs to the engine.[23]

These actions came to fruition on August 26, 1855, when a mill caught fire. Even though the mill was lost and a sixteen-year-old worker killed, Germania No. 1 responded and successfully protected nearby buildings for eight hours. This performance attracted enthusiastic praise from the mill owners and the newspaper.[24]

A year later, in August 1856, Germania No. 1 and its fire engine were taken by ferry across the Fox River to Fort Howard to fight a fire at the Blesch Brewery. The new community of Fort Howard did not yet have a fire company or engine of its own, and again, Germania No. 1 received great public praise.[25] As an indication of stability, Germania No. 1 continued to regularly hold annual meetings, often re-electing many of its officers.[26] The Green Bay Fire Department had achieved both success, and unlike its predecessors, permanence.

About two years after Germania No. 1 organized, GBFD expanded. An editorial in the *Green Bay Advocate*, published fourteen months after Germania No. 1 was formed, stated the city was "getting too large for one engine" and that "when a fire does get under headway here, one engine is utterly inadequate."[27] As further argument for a second engine, the paper stated that because of inadequate fire protection, "it is now almost impossible to get property insured in a reliable company."[28] Another article claimed the need for a second engine company by proposing "honorable rivalry which is so necessary to the support of a fire department."[29] Suggesting that competition between companies was essential for a volunteer fire department to succeed also partly may explain why the three previous solitary fire companies had failed to endure.

The initiative to create a second fire company soon came to realization. A petition presented to the Common Council in August 1856 asked it to purchase a new fire engine. The council promptly authorized the mayor to do just that.[30] Soon after, another petition signed by forty-eight citizens asked the council to recognize a new fire

> *"When a fire does get under headway here, one engine is utterly inadequate."*
>
> - *Green Bay Advocate* editorial

> "Resolved — That the above named persons be authorized to organize themselves into a Fire Company, to be known and designated as the "Guardian Fire Engine Company No. 2 of the City of Green Bay."

Creation of Guardian Fire Engine Company No. 2 from the October 14, 1856, Common Council meeting minutes. This company's lineage can be traced directly to current Green Bay Metro Fire Department Engine No. 2.

company and give it the fire engine just ordered by the city. The council passed a resolution on October 14, 1856, sanctioning "Guardian Fire Engine Company No. 2 of the City of Green Bay."[31] Assuming the nickname "Wide Awake," Guardian No. 2 became the second active GBFD fire company.[32]

Initially, Guardian No. 2 used the Old Croc hand pumper, which the borough had loaned to a mill on the west side of the Fox River.[33] Later, the Common Council purchased a new Button and Blake hand pumper fire engine for Guardian No. 2 in April 1858—one and one-half years later.[34]

A welcoming celebration brought the new Engine No. 2 and existing Engine No. 1 in a parade to a dock on the river for a side-by-side comparison. According to the newspaper, both engines generated impressive hose streams, but there was no clear winner. Competition between companies was openly encouraged. The event culminated with a party for the firefighters.[35]

To further celebrate the christening of the second engine company, a massive and beautiful banner was presented to Guardian No. 2 as part of the city's July Fourth celebration.[36] This banner is currently part of the Neville Public Museum collection. Examination of the existing fire company roll call books (also housed in the Neville collection) reveals Guardian No. 2 was successful and well-supported. These records show that from 1859 to 1863, Guardian No. 2 membership ranged from thirty-eight to sixty-seven members.[37] Given that the hand pumper was physically demanding and required numerous firefighters to operate, Guardian No. 2 was well-staffed. Thus, with the arrival of their fire engine, Guardian No. 2 became a fully active engine company and Green Bay enjoyed an unprecedented level of fire protection.

Another fire company was created shortly after the founding of Guardian No. 2. The Common Council acknowledged in April 1857 that "there are no fire hooks and ladders belonging to the City of Green Bay," and passed

Banner given to Guardian No. 2 in 1858. The center of the emblem is an open eye, symbolizing the company's "wide awake" motto, and is surrounded by interwoven fire hoses. This banner is currently in the Neville Public Museum collection. The photo on page 42 shows the banner in a historical setting.

A. WEISE,
Carriage, Wagon & Cutter Maker,
also dealer in
Carriage Goods, Hubs, Spokes, Etc.
Painting and Repairing of all Kinds done to order.
194 Washington St., - Green Bay, Wis.

Advertisement from the 1872 *City Directory*. In December 1857, Albert Weise built the first wagon used by Washington Hook and Ladder Company No. 1. Over the next two decades, the Common Council frequently paid him to repair various fire department apparatuses. Weise was a founding member of Germania No. 1.

a resolution to purchase such equipment.[38] In addition, on December 26, 1857, the council accepted a hook-and-ladder apparatus made by Albert Weise, a local wagon builder and member of Germania No. 1.[39] At the same meeting, the council accepted a petition signed by seventeen citizens presenting the constitution and bylaws of the Washington Hook and Ladder Company No. 1. The "Hooks" were immediately given possession of the new hook-and-ladder carriage.[40] Workers completed a shed to protect the apparatus a few months later at a cost of $60.08.[41] Further arrangements were made shortly afterward for a larger space to house both the Washington No. 1 and Guardian No. 2 apparatuses.[42] This was the first floor warehouse of the Phillip Klaus store on the south side of Pine Street between Washington and Adams.[43]

By the middle of 1858, GBFD had blossomed into one ladder and two engine companies in two separate stations. At the September 30, 1858, General Review, an alderman reported to the Common Council that the fire department "engines, hooks, ladders, and other apparatus [were] in good order," with ninety-five total members between the Germania No. 1, Guardian No. 2, and Washington No. 1 companies.[44] The Green Bay Fire Department had become extensive, diverse, and this time, enduring.

Fire department ordinance heading from March 27, 1858. The Common Council established comprehensive rules, duties, responsibilities, and operational guidelines for the Green Bay Fire Department. At this point, the department consisted of two engine companies as well as one hook-and-ladder company.

The Green Bay Common Council recognized the need to institute formal regulation and structure for the city's three fire companies. In March 1858, the council passed "An Ordinance to Organize the Fire Department."[45] In great detail, the ordinance defined the roles, duties, and authorities of the chief engineer (in overall command of the fire department; the modern term is fire chief), the assistant chief engineer, officers, and firefighters. The chief engineer and assistant were nominated by the firefighters themselves and confirmed by the Common Council.

The ordinance also addressed problems of the past by emphasizing maintenance of apparatus and equipment. Since the two fire engines were hand pumpers, powered by humans and physically arduous, the chief engineer and assistant had the authority to order bystanders to assist, and even empowered the constable and marshal to arrest those who refused. Those found guilty faced a $5 to

> **New Engine House.**—By reference to the Council proceedings, it will be seen that Engine Co. No. 2 and the Hook and Ladder Co. are to occupy the lower room of P. Klaus & Bro.'s store, on Pine st., which is being fitted up for the purpose.

Fire department news from the May 13, 1858, *Green Bay Advocate*. Until December 1859, the Common Council rented a warehouse room for Guardian No. 2 and Washington No. 1 on the south side of Pine Street, between Washington and Adams. This area was eliminated with construction of the downtown mall in the 1970s.

Guardian Fire Company No. 2 preparing for the July 4, 1870, parade. Firefighters are holding the ropes they used to haul the 1858 Button and Blake manufactured hand pumper (in back) and the hose cart. At a fire, they parked the hand pumper at the water source (river or well) then unrolled the fire attack hose from the reel on the hose cart. Firefighters manually operated the long poles (brakes) on the hand pumper in an up-down motion to power the internal water pump. In this image, two horses are attached to the hand pumper, but normally only firefighters hauled it to a fire. Also, five firefighters (officers and foremen) are holding speaking trumpets, used to convey verbal orders at the busy and loud fire scenes. The banner displayed on the hand pumper in the rightcenter of the photo was given to Guardian No. 2 in 1858 and is currently in the Neville Public Museum's collection (see page 39).

North side of Green Bay in 1876. The above image, taken from the west side of the Fox River, shows the Guardian No. 2 Engine House, built in 1859 (left-center, just beyond a row of trees, and close-up in the photo at right). Note the three-story-high tower on the engine house, from which the fire alarm bell sounded to summon the volunteer firefighters. The top photo looks southeast and shows North Washington (just beyond the boats) and North Adams (one block beyond). The east end of today's Ray Nitschke Memorial Bridge (Main Street) would be in the foreground in front of the boat's bow.

$25 fine. It is unknown how often this was actually done. Upon passage of the ordinance, the mayor stated the fire department was in a "perfect state of organization."[46]

By mid-1858, firefighting in Green Bay had reached an unprecedented level of progress. Two fire companies, each with a hand-pumper fire engine as well as a hook-and-ladder company, were active and governed by a city ordinance. The Green Bay Fire Department thrived during a subsequent period of growth, support, success, and recognition.

The addition of further fire companies soon became one of the clearest indicators of growth. A unique fire department company formed in September 1859 after the Common Council received a petition from thirty citizens asking to be recognized as a branch of the fire department named "Green Bay Sack and Protection Company."[47] The next month, the council passed an ordinance giving this fire company the authority to enter any building threatened by fire in order to remove property and goods.

The Sack Company was fully responsible to "protect and guard" the removed property until owners could take possession. Members had the authority to order bystanders to assist and police powers to arrest anyone caught stealing or looting.[48] This fire company focused exclusively on what modern firefighters would term salvage.

Sack Company was part of the annual fire department inspection in April 1860.[49] Following a downtown fire the next month, the Sack Company was specifically complimented for saving the personal property of a family, as well as the stock and furniture of a business.[50] In May 1861, the company's apparatus was again part of the annual fire department inspection.⁵¹ However, the organization soon ceased to exist, as a May 1863 fire department inspection report did not list Sack Company.⁵² The reasons for its demise are not known. Although this unique fire company lasted a few years at most, the Sack Company stands as a marker for the experimental development of GBFD.

Another engine company formed about the same time as Sack Company. In February 1860, the Common Council received a petition from a group of citizens asking to recognize a new engine company on the far side of the East River, which was then the eastern edge of the expanding city. At the next meeting, March 10, 1860, Franklin Engine Company No. 3 officially was accepted and given Old Croc, the first fire engine bought from the military in 1843, along with a hose cart and hose.⁵³ Although fairly old and dilapidated, a special committee had previously examined Old Croc while in storage at a local warehouse and found it "in tolerable good order."⁵⁴ With the creation of Franklin No. 3 in early 1860, GBFD now had three engine companies.

As a showcase, a Firemen's Tournament was held at the Brown County Fair that year, pitting the three Green Bay engine companies and the newly formed Borough of

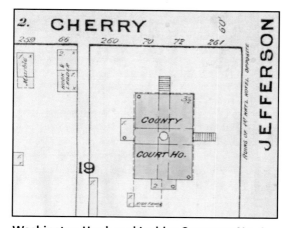

Washington Hook and Ladder Company No. 1 house from the 1879 Sanborn Insurance map. This site (just under the "66" in the upper left of the image) is on the south side of Cherry Street between North Adams and North Jefferson. It is currently a parking lot. The nearby County Court House site is the southwest corner of Jefferson and Cherry, which is today the Brown County Human Services Department building and parking lot.

Fort Howard engine company, with the winner determined by the longest reach of a hose stream. Guardian No. 2 won with 188 feet and Franklin No. 3 came in last at 140 feet. Guardian No. 2 was awarded a magnificent silver commemorative trumpet that is part of the Neville Public Museum collection.[55] This competition highlighted the great advancement in firefighting in Green Bay. Whereas less than six years earlier there were no active fire companies, this competition featured four separate engine companies. The area had seen a drastic improvement in firefighting.

Expansion of the fire department was a great indicator of the prosperity of Green Bay, with the April 1860 inspection parade featuring five companies: Germania No. 1, Guardian No. 2, Franklin No. 3, Washington No. 1, and Sack and Protection.[56] Although change continued, systematic firefighting in Green Bay was now permanent.

Along with growth, the blossoming GBFD received substantial financial support. The city paid $175 per year rent for a warehouse on Pine between Washington and Adams for Guardian No. 2 and Washington No. 1.[57] Then, in December 1859, both Guardian No. 2 and Washington No. 1 moved to a newly built fire station, paid for by the Common Council, on the east side of North Adams, one and one-half blocks north of Pine.[58] However, within half

Firemen's Tournament results from the September 27, 1860, *Green Bay Advocate*. Three Green Bay and one Fort Howard hand pumper engine companies competed against each other at the Brown County Fair. This is the only recorded instance of this sort of formal competition. To commemorate their victory, Guardian No. 2 received this ornate, ceremonial silver trumpet (below), which is in the Neville Public Museum collection.

a year, conflict and crowding caused Washington No. 1 members to move to their own station, which they built at their own expense on the south side of Cherry Street between Adams and Jefferson.[59] Guardian No. 2 remained at the engine house on North Adams.

About the same time, the Common Council provided funds to build an engine house for the new Franklin No. 3 company.[60] This simple wooden building, partially paid for by the Franklin No. 3 firefighters, was located on the south side of Main Street at the intersection with modern North Irwin Avenue.[61] By the mid-1860s, there were four fire stations in Green Bay. GBFD had truly expanded.

Yet with expansion came increased costs for equipment and maintenance, as well as other needs. The Common Council would consider each individual request for funds, almost always resulting in further financial support for the burgeoning GBFD. For example, shortly after forming, Germania No. 1 was granted $1 per day in the winter to maintain a fire at the engine house and protect the fire engine from freezing – a well-known peril in Northeast Wisconsin.[62]

Each of the new stations was also heated at the borough's expense. Starting in 1861, when there were three

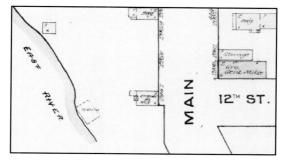

Franklin No. 3 engine house from the 1883 Sanborn Insurance map. The engine house (lower-center on the image) was on the south side of Main Street at the intersection with 12th Street (modern Irwin Avenue). The platform adjacent to the East River behind the engine house likely served as a water supply access point for the hand pumper.

Dawn of the Green Bay Fire Department

engines in separate stations, the city paid individuals to maintain these warming fires. They held the additional duty of maintaining access holes through the ice for fire department use.[63] By 1862, these duties transferred to the city night watchmen.[64] In November 1861, the Common Council paid to have a dock built on the East River at Irwin large enough to fit two engines for simultaneous river water use.[65]

Germania No. 1 firefighters reported in April 1959 that at a recent drill, they discovered their hemp hose was worthless and requested "double riveted leather hose."[66] Within a week, a special committee of aldermen confirmed this issue and also found some deficient hose being used by Guardian No. 2. They immediately recommended purchasing 250 feet for each company, specifying "the best kind of leather hose."[67] Although the fire pumpers were hand-drawn by the firefighters, the Common Council often paid horse teams to haul the fire engines, especially through deep snow.[68]

Because the volunteers responded from home or work, it was necessary to develop an alert system. In 1856, a bell was provided for Engine House No. 1 on South Washington.[69] A few years later in 1863, the Common Council purchased bells for Engine House No. 2 and No. 3, and the hook-and-ladder station.[70]

A singularly telling event reveals the commitment from the Common Council toward the growing GBFD. With new companies, engine houses, and equipment needs, the demands on the Common Council grew to the point that on June 1, 1861, a permanent Fire Department Committee of three aldermen formed to contend with the numerous — and increasing — fire department issues and requests.[71] Some evidence suggests the first three attempts at fire companies in Green Bay failed in part due to lack of financial support. From 1854 onward, this support was forthcoming, generous, and unquestionably facilitated the permanence of the Green Bay Fire Department.

Not only did the developing GBFD enjoy financial support from city authorities, but the citizens and newspaper were absolutely supportive. A parade on New Year's Eve, 1855, celebrated the one-year anniversary of Germania No. 1. Afterward, the *Green Bay Advocate* commended the popularity of the new engine company.[72] Similarly, Guardian No. 2 had a party in December 1856, shortly after forming, and a second party in February 1858.[73] Franklin No. 3 and Washington No. 1 soon held parties as

Fire warden ordinance heading from the City of Green Bay ordinance book. The Common Council passed this ordinance November 25, 1854, which reaffirmed an 1842 ordinance giving the fire wardens the authority to inspect every occupied building in the borough. The only construction rules in the ordinance vaguely addressed stoves, fire places, and associated pipes and chimneys—common causes of fires.

Official notice from the November 22, 1855, *Green Bay Advocate*. This was the first public use of the term "Green Bay Fire Department." The chief warden, later known as chief engineer, was in overall command of the department and is equivalent to the modern term fire chief.

well. These functions served not only as social gatherings, but also as fundraisers. In fact, Klaus Hall, located on the south side of Pine between Washington and Adams, was used so frequently for fire company events that it became known as Firemen's Hall until, ironically, it was destroyed by fire in 1863.[74]

As required by ordinance, the fire department held the semi-annual fire department inspection, which also became a social event. The fire companies marched in parades and displayed their prowess, generating confidence in the emergent GBFD.

Along with progress on the suppression side of firefighting, prevention of out-of-control fires was also underway. The series of significant fires in 1853 and 1854 that led to creation of the city's first fire companies also triggered changes to fire prevention. About the time Germania No. 1 was being formed, the Green Bay Common Council passed Ordinance No. 18 on November 25, 1854, re-affirming and expanding the duties of the fire wardens.[75] The ordinance authorized a chief warden and two assistants in each of the North and South wards. In fact, fire wardens actually became paid positions. New laws featured strict specifications regarding stove pipe construction, a frequent cause of fires, and the fire warden ordinance was updat-

Washington Street in 1857. Fire easily spread between the all wooden-exterior buildings. This hazard prompted the fire limits ordinance of November 1859, which permitted construction of only stone or masonry exteriors in the congested downtown area. In this image, there are at least two brick-exterior buildings, built even before the ordinance passed.

ed in 1858 with even more details on required stove pipe construction methods.[76] Perhaps most indicative of the city's attention to fire prevention came in 1854, when the Common Council diligently began assigning fire wardens, especially when a vacancy occurred.[77] Fire prevention was not allowed to lapse.

The consistent presence of fire wardens paid off in measurable fire prevention statistics. A story in the February 18, 1858, *Green Bay Advocate* noted the absence of recent serious fires. The paper credited citizen diligence and especially the efforts of the fire wardens. Similarly, a year later, the paper reported, "Our city has been remarkably exempt from disasters for two years or more."[78] The paper often encouraged safe disposal of ashes and promoted maintenance of stoves, chimneys, and fireplaces. Although fires did continue to occur in Green Bay, fire prevention had been established.

The passage of the first building construction laws in Green Bay solidified further advancements in fire prevention. A contributing factor to the destructive scope of the 1840, 1841, 1853, and 1854 downtown fires was the exclusive use of wood for construction material, particularly for building exteriors. Recognizing this issue in his April 1856 inaugural address, Mayor F. E. Eastman specifically suggested regulating building material in the downtown mercantile district so as to avoid repetition of those devastating fires.[79]

In November 1859, the Common Council passed "An Ordinance prescribing fire limits and construction of buildings therein."[80] Fire limits were a defined area, or zone, within which there were specific construction requirements and restrictions. The fire limits were set around the business district, which was the eleven full

Fire limits ordinance heading from the November 1, 1859, Common Council minutes. This ordinance established the first extensive construction requirements in Green Bay. Applied exclusively to the congested downtown, this ordinance allowed construction of only masonry or stone exteriors. Inherently, this prohibited wooden exteriors, though existing buildings were exempt.

or partial blocks along Washington Street, extending south to north from Doty Street to the East River.[81] This area had become a true commercial center, with numerous tightly packed buildings, and as such, a great potential for conflagration. Within the fire limits, all exterior walls had to be stone, brick, or other fireproof material, while gutters and roofs had to be metal. End walls and party walls (between two abutting buildings) were required to extend at least thirty inches above the roofline. Though not formally stated, inherent to these rules was the prohibition of wood exteriors. Addressing another frequent cause of fire, waste ash was to be deposited in receptacles made of brick or similar material. Failure to comply resulted in fines set at $20 to $100 per week.[82] A later modification filled a loophole by prohibiting property owners from moving wooden buildings into the fire limit's area.[83]

Worth note is the fact there were no restrictions on building materials outside the downtown (fire limits) area. Restrictions still existed for heating and cooking appliance construction, as had been established earlier. Even before the fire limits were set, a "fire-proof" building,

> Fire—Almost.—On Saturday last, about 1 o'clock, the rear part of Mr. C. J. BENDER's cabinet shop was discovered to be on fire. The citizens and firemen were instantly on the spot, and the fire was extinguished with (riding injury to the premises. Mr. B. requests us to tender his thanks to the citizens and Germania Fire Co. No. 1, for the prompt and timely service rendered.

Thank you from business owner in the July 23, 1857, *Green Bay Advocate*. Firefighters often received acknowledgement in the newspapers. In this case, the business owner also thanked citizens, most likely for assisting with operation of the hand pumper brakes, a physically demanding job.

> **Fire Department—***Officers Elected.*—The following are the officers of the several companies for the ensuing year, as far as we can ascertain.
>
> *Chief Engineer.*—Anton Klaus.
> *Assistant Engineer.*—C. O. Lovett.
> *Treasurer.*—C. L. Wheelock.
>
>   Germania Engine Co. No. 1.
> *Foreman.*—Gustave Huinker.
> *Assistant.*—Frank Lenz.
> *Hose Captain.*—Philip Frank.
> *Assistant.*—Anton Basche.
> *Treasurer.*—Chas. Klaus.
> *Secretary.*—Fred. Burkard.
> *Engineer.*—Ernst Strauble.
> *Assistant Engineer.*—Christoph Meister.
>
>   Guardian Engine Co. No. 2.
> *Foreman.*—L. J. Day.
> *Assistant.*—P. I. Earle.
> *Hose Captain.*—C. D. Suydam.
> *Assistant.*—Frank Boyd.
> *Secretary.*—J. Kip Anderson.
> *Treasurer.*—C. L. Wheelock.
> *Engineer.*—Abraham Lucas.
>
>   No. 3—*Not reported.*
>
>   Hook and Ladder No. 1.
> *Foreman.*—Louis Scheller.
> *Assistant.*—Jean.
> *Secretary.*—J. A. Killian.
> *Treasurer.*—Martin Huber.

Fire department election announcement from the January 22, 1863, *Green Bay Advocate*. The firefighters nominated individuals for overall fire department positions (chief engineer, assistant, and treasurer) who were then confirmed by the Common Council. Members of each company selected their own officers.

probably the first in Green Bay, was constructed in 1856 at the southwest corner of Washington and Pine.[84]

From today's perspective, the best measure of success for the newly developed GBFD came in the form of community response and newspaper reports. Immediately after its 1854 inception, Germania No. 1 began to receive newspaper praise. As more fire companies formed, the newspapers continued to lavish commendations on the firefighters and GBFD. Some examples found in these early years are as follows:

- The "firemen were promptly on hand" preventing spread to adjacent homes.[85]

- "To Germania Fire Company No. 1. — The subscriber would hereby tender his thanks to said Company for their timely and efficient service at the burning of his Shingle factory on the morning of Friday last. J. INGALLS."[86]

- "The several Fire Companies were on hand with commendable alacrity."[87]

- To the Common Council: "Cherish the fire department of our city as you would cherish your dearest interests."[88]

- "Had it not been for the directed exertions of a well-organized fire department ... [the fire] would have been a very serious affair."[89]

- The "United States Hotel was in great danger, but through the exertions of the firemen was saved."[90]

- But for the fire department, the fire "might have laid in ashes the principal business portion of our city."[91]

- "The advantages of a good, well-organized and efficient fire department were well-displayed."[92]

- "Our Fire Department cannot have too much praise."[93]

- Praise for Franklin No. 3 responding to a fire downtown and "dragging their machine through the mud the whole distance."[94]

Without a doubt, the new GBFD was developing into a well-appreciated organization.

Through the end of 1863, GBFD had advanced tremendously from its faltered beginnings. Although fire was always an obvious danger, several catastrophic blazes occurred before municipal authorities and citizens had taken sustained action. The April 21, 1859, *Green Bay Advocate*

Members of Guardian No. 2 about 1860. The "Hose 2" on the caps indicates these firefighters were members of the hose company, rather than the engine company. These were separate companies, but still part of Guardian No. 2.

objected to a proposal to temporarily discontinue the night watchman by arguing that "probably no city in the west has suffered more from severe fires than Green Bay."

Organized firefighting bloomed in Green Bay once Germania No. 1 provided a firm foundation. More fire companies were established, more engine houses built, and more equipment needs fulfilled through Common Council funding. Most significantly, the response from the community was overwhelmingly supportive. GBFD was providing exceptional service.

However, a disastrous fire – one larger than any seen before – would test this new department. Such a challenging event would lead to the next major force of change.

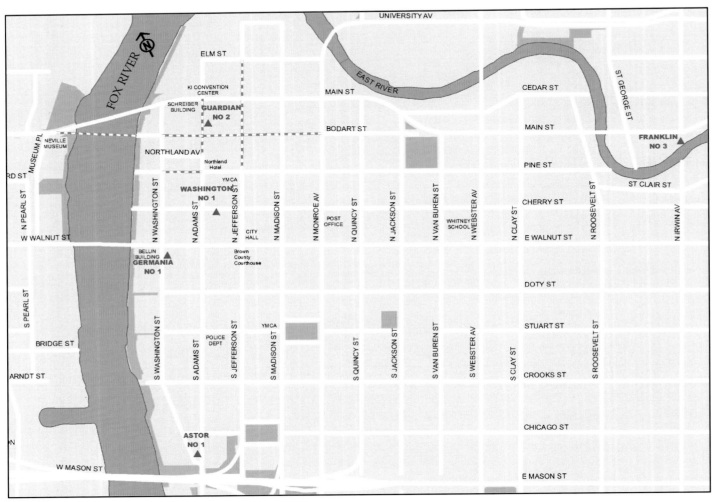

GBFD fire station locations (1863-1875) within modern Green Bay. Fire stations discussed in this chapter are indicated by the triangles and labeled with the company names. Streets that no longer exist as of 2016 are indicated by dashed lines.

Chapter 4

Transition to Steam Fire Engines
1863 - 1875

The Green Bay Fire Department (GBFD) consisted of three hand pumper engine companies and one hook-and-ladder company by the end of 1863.[1] With the first bridge across the Fox River completed that fall, the Fort Howard Fire Department could respond quickly across the river, providing another hand pumper and a contingent of firefighters. GBFD frequently responded into Fort Howard as well. This act of reciprocal assistance, today known as mutual aid, in reality provided up to five fire companies on both sides of the river, very notable for a combined population of under 3,000.[2]

Reliance on the hand pumpers revealed a weakness in this otherwise impressive situation. Humans provided the power to the water pump. Long handles, known as

The flames, fanned by the wind, spread from building to building through the congested commercial area.

brakes, were pulled up and down by firefighters, moving a series of levers that ultimately provided energy to the water pump mechanism. Twelve to twenty firefighters were required to simultaneously work the brakes, depending on the size of the hand pumper. An individual could last five to ten minutes before needing to rest. Since some operated the brakes while others rested, it was necessary to have a large number of firefighters available. Bystanders occasionally were needed, and fire officials had the legal authority to compel assistance. However, sometimes not enough people were available.

Officials recognized the issue of manpower needs for the hand pumpers as far back as 1854. The first fire warden ordinance established the authority for them to force "idle bystanders" to assist on the brakes. Failure to comply resulted in arrest and a $5 to $25 fine.[3] It is unknown how often, or even if, bystanders were actually ordered to assist and if those refusing were arrested.

A series of fires during the early and mid-1860s dramatically revealed the problem with relying on human power. First, around 2 am on November 11, 1863, a fire began in a building at the northwest corner of the intersection North Washington and Cherry streets. The flames made "considerable headway before notice" at this

The Fire of November 11, 1863

Headlines from the November 12, 1863, *Daily Milwaukee News*. Conflagrations often struck US cities in the 1800s. Milwaukee experienced at least five between 1849 and 1892. The most devastating occurred on October 28, 1892, in the Third Ward, destroying 440 buildings over sixteen city blocks.

Local headlines. From the November 12, 1863, *Green Bay Advocate.* (above) and the November 14, 1863, *Appleton Crescent* (below). The worst fire to date in Green Bay destroyed thirty buildings downtown, all within the block encompassed by Washington, Adams, Cherry, and Pine streets. The massive fire overwhelmed the capabilities of the hand pumper fire engines.

early hour, and a moderate wind from the south quickly spread the fire to adjacent structures.[4] Nearly all the nearby buildings had wooden, flammable exteriors. The fire limits ordinance mandating fire-proof exteriors had been enacted four years before, but this only applied to new construction. Pre-existing buildings were excluded.[5] The flames, fanned by the wind, spread from building to building through the congested commercial area.

Green Bay and Fort Howard fire departments quickly responded. All four hand pumper fire engines were on the scene, but firefighters could not extinguish the flames. The inferno "swept through" the entire block bounded by

Transition to Steam Fire Engines

> *"The fire stopped for want of fuel" where the streets acted as urban firebreaks.*

Washington, Adams, Cherry and Pine streets, consuming thirty buildings.[6] Essentially the entire block, nearly two acres, was destroyed "in the heart of the business portion of our city."[7] Efforts by firefighters and citizens to stem the firestorm were futile and "the fire stopped for want of fuel" where the streets acted as urban firebreaks.[8]

Within the burned block, only "Kip Anderson's new fire proof building" and another "patent brick" building survived.[9] Both were built after implementation of the fire limits construction requirements.[10] Many small business were lost, as were the post office, the large US Hotel, and Klaus Hall, also known as Firemen's Hall because of frequent use for fire company social events.

Some buildings across the street did suffer light damage and all were in danger of catching fire. A stronger wind would have spread the fire across Adams or Pine, extending the destruction to the East River, many blocks distant.[11]

All four hand pumpers from Green Bay and Fort Howard operated for hours, and newspaper reports made clear the firefighters were overwhelmed. Other members of the community helped, as occurred during the large 1853 fire. A bucket brigade saved a house across the street from the burning block, stopping the spread of the flames at that location. Goods from several businesses were removed to safety, both in the affected block and where the flames threatened to cross the streets. Citizens accomplished this task, because all the firefighters were committed to working the hand pumpers and hoses.

Operating the hand pumpers for hours exhausted many firefighters.[12] Citizen help on the hand pumper brakes was absolutely necessary. The November 12, 1863, *Green Bay Advocate* complimented some citizens for assisting the firefighters, specifically mentioning that "several ladies took hold of the brakes" of the hand pumpers. Identifying three by name, Mary Joyce, Josephine Forsythe and Miss Rowbotham, the paper stated their "names should be recorded in gold letters."

In contrast to these compliments, the newspaper reprimanded others. The *Green Bay Advocate* pointed out that some "citizens generally worked on the engines with a will, but there were exceptions, as there will always be," and that there was "refusal or neglect of some men to work."[13] A major weakness of the hand pumpers was evident—the requirement for so many people. The number of firefighters and bystanders available, capable, and willing was limited, and in reality, inadequate for such a major fire.

> *"Several ladies took hold of the brakes" of the hand pumpers.*
> - Green Bay Advocate report

Problems with manpower needs of the hand pumpers continued to occur after the November 1863 conflagration. The very next month, the hand pumpers operated for up to five hours at a mill fire in Fort Howard, an incredible demand on those providing the human power.[14] A February 1864 newspaper account describes how the firefighters had to drag the hand pumpers through the deep snow, then were expected to operate the brakes. The paper openly argued that this was too much to expect from the volunteers.[15] The paper praised the firefighters after many ruined their best clothes fighting a fire on a Sunday. But at that same fire, there was "a class of citizens who always stand idly by refusing to work."[16] Idle bystanders were described as "warts on the community," comparable to cowards on the battlefield of war.[17] Municipal ordinances gave the mayor, aldermen, chief engineer, and assistant chief engineer authority to "order and direct any bystanders, or bystanders, who may appear to them as idle spectators at the fire, to assist in putting out such fire, or saving property or life." Failure to obey could result in arrest and fine.[18] This threat was not enough to convince some citizens to join in the efforts at fires.

The issue of inadequate manpower came to a crisis in August 1866. A fire occurred in a large, multi-story

Steamer fire engines in operation. Though not GBFD fire engines, this image provides a good example of steamers in use. The engineer shoveled coal (piled in the street) into the burn box at the base of the boiler. A steamer needed only the one engineer, in contrast to a hand pumper, which required scores of firefighters to function.

warehouse, then rapidly spread to another nearby warehouse and dock, all at the northern end of Washington Street at the river. The fire departments from Green Bay and Fort Howard responded, and again the hand pumpers operated for hours. Several bystanders were arrested for not complying with orders to assist on the hand pumper brakes. It was reported that some had just been relieved from the brakes. Unfortunately, at least one overzealous alderman made them immediately return to work while

"Who will circulate a petition asking the Common Council to purchase a steamer fire engine?"

— *Green Bay Advocate*

others refused and were arrested.[19] This caused hard feelings in the community and among the firefighters. An alternative was needed. Fortunately, advancing technology provided a new type of fire engine.

The solution to the massive human demands of the hand pumper was the steam fire engine, more simply known as a steamer. These fire engines were powered by steam generated from an onboard coal-fired burn box and boiler. The steam drove the mechanical system that ultimately powered a water pump. The resulting water streams had great volume and reach. Similar to the hand pumpers, a short supply hose from the steamer was placed into water sources such as rivers, wells or small, private underground cisterns. Most importantly, the steamers required only a few firefighters to operate. Furthermore, long lengths of hose could be deployed from the steamer, reaching fires well distant from the water sources. Overall, the steamers were an incredible improvement in firefighting capabilities.

The Cincinnati (Ohio) Fire Department put the first steamer into service in 1852.[20] Steamers then began replacing hand pumpers in larger American cities. Ironically, the steamer itself was occasionally a fire hazard. When fully engaged, the coal burner boiler would emit burning cinders with the exhaust. There were instances in Green Bay when these burning cinders actually landed on nearby buildings and caused additional damage.[21]

City officials first considered obtaining a steamer for GBFD prior to the large warehouse fire in August 1866. However, in his April 1866 inaugural address, Mayor Charles Robinson discounted the need for one because of expense and confidence in the fire department's hand pumpers.[22] A few months later, the Common Council actually received an offer from a steamer manufacturing company, but no action was taken.[23] Again, the price was considered too great.[24]

In contrast to the municipal authorities, the *Green Bay Advocate* openly asked, as part of a story on the large August 1866 warehouse fire, "Who will circulate a petition asking the Common Council to purchase a steamer fire engine?"[25] Another article emphasized that because of the low water pressure generated by the hand pumpers, firefighters could not attack fires four or five blocks from the river. A steamer would solve this problem and "may be the means of saving a conflagration."[26] The newspaper decidedly endorsed purchasing a steamer fire engine.

Progress occurred slowly. In May 1867, citizens pre-

Germania No.1 steamer fire engine in 1875. The Amoskeag Manufacturing Company of Manchester, New Hampshire, manufactured this steamer, formally named "Enterprise." The boiler is the large, upright barrel with the smoke stack at the top. Riding at the back is the engineer, who fed coal into the burn box at the base of the boiler. Behind the driver is the steam-driven water pump. A section of rigid supply hose is above the wheels, and another is on the opposite side. Firefighters connected these hose sections, attached one end to the pump inlet and lowered the other into the water source, such as a river or a cistern.

sented petitions to the Common Council, not just in support of the purchase of a steamer, but also against.[27] As a result, a special committee from the Common Council was formed to examine the issue, marking the first formal municipal action.[28] They inspected steamers already in service with Oshkosh and Fond du Lac fire departments in December 1867.[29] The special committee soon recommended the city purchase a steamer manufactured by Amoskeag.[30] Within a few months, the Common Council approved the purchase and passed an ordinance to issue bonds to raise the funds.[31] One newspaper touted that when the steamer arrives, "We will be in good shape to hold the destructive element which has visited us often of late, at bay."[32] Finally, newly elected Mayor Anton Klaus endorsed the prompt purchase of a steamer during his 1868 inaugural address, a reversal from Charles Robinson's position two years before.[33]

The steamer arrived in Green Bay at the end of October 1868. This was nearly five years after the massive and devastating conflagration that dramatically highlighted the shortcomings of the hand pumpers.[34] Manufactured by Amoskeag of Manchester, New Hampshire, this model was a second class, referring to a pump capacity of 700 gallons per minute.[35] The steamer was christened "Enterprise" and given to Germania No. 1, who took on the new name Steam Fire Co. No. 1. They were still referred to as Germania No. 1 and maintained the "Rough and Ready" motto.[36] The city also purchased 1,000 feet of new hose and a hose cart to take advantage of the tremendous pumping capacity of the steamer.[37] GBFD could now fight fires well distant from the rivers, the most frequent water supply source. This was a great improvement over the hand pumpers.

Firefighters quickly held a public trial of the Enterprise No. 1 steamer, thrilling the large crowd in attendance and completely satisfying the citizens, municipal authorities, and fire department members.[38] In one demonstration, firefighters used the steamer to throw a water stream over the tallest building in Green Bay, far outperforming the hand pumpers.[39] Of great significance, this was done with only a small group of firefighters. One or two operated the steamer, while only a few more held the hoses and nozzles. With improved pumping capabilities and a massive decrease in manpower needs, GBFD realized an incredible improvement in firefighting capabilities with this single purchase.

Obtaining a steamer heralded several additional and noteworthy changes to GBFD. First, the mechanical complication of the steamer, compared to the hand pumper, required much greater operator attention. James Lucas was hired as the full-time, paid Engineer of the Steamer, shortly after Enterprise No. 1 steamer arrived.[40] Lucas became the first full-time member of GBFD with a salary of $1,000 per year.[41]

Lucas's work schedule cannot be determined from existing records, but it appears he spent nearly all of his time at or very near the fire station. The first city directory (1874) actually lists the residence of the Engineer of the Steamer as Engine House No. 1. His home was the fire station.[42] This position was maintained, though the salary was reduced to $800 in 1869 and $700 in 1871. The salary eventually settled at $500 through the early 1880s, then rose again to $625 in the mid-1880s as responsibilities expanded.[43]

The large size of the steamer made construction of a new fire station necessary. Engine House No. 1, built about 1854 on South Washington Street just south of Walnut, could not accommodate the twenty-four-foot-long, six-foot-wide, and nearly nine-foot-tall steamer.[44] To make room for a new, larger engine house, the existing two-story, wooden engine house was moved in the fall of 1868 to the southern end of the triangular park at the intersection of Washington and Adams streets.[45] This is the same block as current Fire Station No. 1, which sits on the northern end of that small park.

A single-bay, two-story engine house was built on the now vacant lot on Washington. New Engine House No. 1 opened in January 1869. It featured a brick exterior, a bell tower in front, and a sixty-two-foot-tall hose tower in the rear.[46]

An Amoskeag steamer fire engine, most likely GBFD Enterprise No. 1. A pile of coal is behind the bin, ready to be shoveled into the burner. The larger hose to the lower right is the rigid supply line, most likely draped over the side of the dock into the river. Next to it is the smaller attack hose, extended to the burning building.

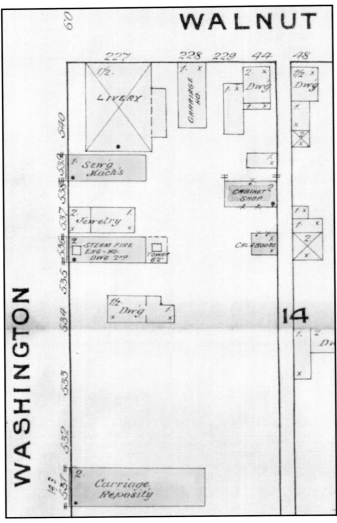

Germania No. 1 Steamer Engine House from the 1879 Sanborn Insurance map. The city built this two-story, single bay station (left-center above) in 1868 on South Washington to accommodate the larger Enterprise No. 1 steamer. It featured a bell tower in front, and hose drying tower in back. This site is currently the southern end of the Backstage at the Meyer. Behind the engine house, the blue structure labeled "calaboose" is the small, rudimentary city jail.

With the Enterprise No. 1 steamer now operated by Germania No. 1, the Smith hand pumper, purchased in 1851, was no longer being used, at least briefly. On December 26, 1868, the Common Council recognized Astor Fire Co. No. 1.[47] The city gave the new company the Smith hand pumper and allowed it to occupy the engine house recently moved to the park at Washington and Adams.[48] This greatly improved fire protection in the expanding South Ward of Green Bay.

GBFD also made changes to improve its water supply. Firefighters had utilized the Fox and East rivers, but the hand pumpers could only push water through the hoses a few hundred feet at most. Beyond that distance, they used small, private wells near the burning building, though the volume of water available was always inadequate.

The need for substantial and reliable water sources in neighborhoods away from the river became apparent. In March 1868, before the steamer arrived, a house four blocks from the Fox River was entirely destroyed by fire, principally because of the lack of water nearby.[49] This was an unfortunate, but common problem on the edges of the growing city.

Another consideration was the volume of water pumped by the steamer was much greater than that from the hand pumpers. While the rivers offered an inexhaustible supply, the small domestic-use wells in the neighborhoods away from the rivers were completely inadequate to support the 700 gallons-per-minute capacity of Enterprise No. 1. The wells would empty, rapidly leaving firefighting efforts ineffective.

The solution to both issues (distance from the river and water volume) was construction of large, underground tanks, also known as cisterns. The same Common Council committee that sought the steamer also recommended the construction of six cisterns.[50] The mayor endorsed construction of cisterns, and the costs were included with the steamer bond.[51] Accordingly, two cisterns were built under Madison Street, four blocks from the Fox River, soon after Enterprise No. 1 steamer arrived. These interconnected cisterns each contained over 20,000 gallons of water and were fed by a spring.[52] The cisterns tremendously improved water supply in the expanding east side of Green Bay.

A last change came about due to the size of the steamer itself. Enterprise No. 1 steamer was comprised

Wooden water pipes. Green Bay Water Utility workers found these approximately six-foot long wooden pipes during street excavations. Though the original use is not known, these would be similar to the wooden pipes used to interconnect underground cisterns. When the steamer fire engine drew water from one cistern, water flowed from other cisterns many blocks away through wooden pipes, vastly increasing the total volume available. Close-ups of the tapered and grooved ends show how the sections were joined, then most likely sealed. These pipes could withstand only a few pounds of pressure without leaking.

mostly of metal and had water within the onboard boiler, at 7,000 pounds making it substantially heavier than the hand pumpers.[53] While the hand pumper could be challenging for the firefighters to haul by hand, moving the steamer was extremely difficult, if not impossible, especially when wet weather turned the dirt streets of the era into impassable mud paths.

The obvious solution was to haul the steamer by

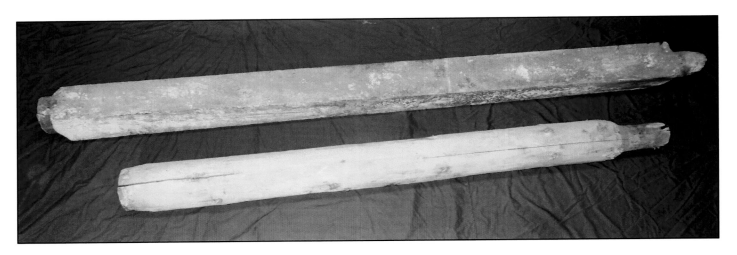

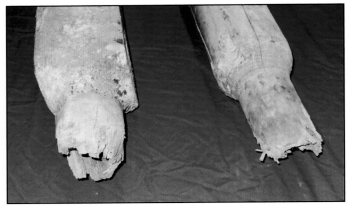

horses. However, the cost of purchasing a horse team exclusively for the steamer was prohibitive. The Common Council's solution was to implement a "bounty" system in which it gave $10 to the first horse team to respond after an alarm sounded and $5 to the second team to respond. Payment was dependent on returning the steamer to the engine house after the fire.[54] The hose cart carrying 1,000 feet of hose purchased along with the Enterprise steamer weighed 4,600 pounds, so the city also applied the horse team bounty system to the hose cart.[55] The Fire Department Committee reported bills for $10 and $5 very shortly thereafter.[56]

Although the bounty system met with much success, there were problems. After Enterprise No. 1 steamer arrived late to a fire, the *Green Bay Advocate* urged that the city should own a "good, strong team of horses, kept

The Smith hand pumper in the possession of the Algoma (Wisconsin) Fire and Rescue Association in 2015. Germania No. 1 began to use the Enterprise steamer, so the city gave the Smith hand pumper to the newly formed Astor Fire Company No. 1 in 1868. This company's marking on the upright metal air-chamber tower (above) has been retouched.

(Opposite) Fire alarm bell from GBFD Engine House No. 1 currently at St. Agnes Church in Amberg, Wisconsin. GBFD used this approximately 3,500-pound bell to sound alarms at the South Washington engine house from 1874 until removal in about 1925. The bell is embossed with "The Jones & Company Troy Bell Foundry, Troy, N. Y., 1874," which matches exactly with the Green Bay Common Council records. This bell remained in the Station 1 tower until 1925, when it was donated to St. Joseph Church in Green Bay. In 1984, it was given to St. Agnes Church in Amberg. Today, it sits in a tower in front of the church and rings at noon and to announce service.[63]

convenient to the steamer."[57] The Common Council briefly considered this option, but rejected it as too expensive.[58] The bounty system remained in place.

By early 1869, GBFD had undergone several years of profound change. A new steamer was purchased and placed in a new engine house. The previous engine house and hand pumper were moved to the South Ward of the city and now operated by a new fire company, Astor No. 1. The first full-time, paid GBFD firefighter, the Engineer of the Steamer, had been hired, and very large cisterns for water supply were constructed in the expanding east side. These changes were consequences of the 1863 conflagration and other major fires, all of which highlighted the weaknesses of the hand pumpers. GBFD now consisted of Steam Co. No. 1, hand pumper Guardian No. 2, Franklin No. 3, Astor No. 1, and the Washington No. 1 Hook and Ladder Company. An ordinance passed a few months after the steamer arrived reaffirmed and expanded fire department roles and responsibilities.[59] In just a couple of years, GBFD had emerged from rudimentary status to become a well-developed fire department.

There were many minor changes to GBFD over the next few years. A 2,500-pound bell was placed in the tower of the new Engine House No. 1 on South Washington.[60]

This was a great improvement over the previous bell. It was so good that the Engineer of the Steamer rang the bell at set times every day (7 am, 12 noon, 6 pm) for the benefit of the public.[61] The bell was recast in 1874, adding 900 pounds and necessitating an expansion of the bell tower to accommodate it.[62]

The open tower provided better sound and enabled citizens throughout the city to hear this big, loud bell.[64] The fire department then implemented a new system for fire alarm notification. After sounding an "initial general alarm," the bell would sound one strike for a fire in the First Ward, two strikes for the Second Ward, and three for the Third Ward.[65] Thus, the five GBFD companies, each responding from different stations, knew the approximate location of the fire.

Enterprise No. 1 steamer was clearly a successful addition to GBFD, so much so that the city soon purchased another steamer. The Fire Department Committee of the Common Council solicited bids for a second steamer less than three years later.[66] However, no action was taken for over a year. This delay prompted citizens and firefighters to send a petition to the Common Council requesting immediate action toward purchasing a second steamer. In fact, the petitioners pledged to make the first payment.

The Common Council quickly heeded the request, but decided the city would make all payments. Within a month, it contracted with Clapp and Jones of Hudson, New York, for a new steamer.[67]

The new steamer arrived in Green Bay at the end of August 1872.[68] It was formally named "Guardian No. 2" and given to that fire company.[69] Slightly smaller than Enterprise No. 1 steamer, the Guardian No. 2 steamer featured mounted lamps with the emblem and "Wide Awake" motto of that fire company.[70] Guardian No. 2 steamer threw a water stream 250 feet straight into the air during the acceptance trial. It was a very capable fire engine.[71] To take advantage of this capacity, the city purchased 1,000 feet of additional hose for Guardian No. 2.[72] The city obtained a large, four-wheeled hose cart within a few years to better transport the considerable amount of hose now available.[73] As with the first steamer, a second full-time, paid Engineer of the Steamer was hired for Guardian No. 2.[74] By the end of summer 1872, GBFD had two steamers in service, both manned by full-time engineers.

The hand pumper fire companies continued to be part of GBFD. In fact, the city purchased a used, but reportedly nearly new, hand pumper in 1870 for Franklin No. 3.[75] This fire company, formed in 1860, had been using

> **The New Steam Fire Engine.**—The new steamer arrived on Saturday last and was deposited in Guardian No. 2 house, to which Company it had been previously assigned by the Council. It is a really elegant piece of mechanism, and has, since its arrival here, held a sort of levee in its new quarters, large numbers of our citizens calling to see it.

Arrival of Green Bay's second steamer celebrated in the August 29, 1872, *Green Bay Advocate*. In contrast to Germania No. 1's steamer "Enterprise," the Guardian No. 2 steamer was simply and inelegantly named, "Guardian No. 2," which is on a plaque affixed to the side of the boiler. The newspapers frequently reported fire department news.

the Old Croc hand pumper, first obtained by Green Bay from the military in 1843.

Other changes occurred in Green Bay in the early 1870s. Several buildings in the expanding east edge of the city were lost to fire because they were too far from either the new cisterns or the two rivers.[76] In response, three additional underground cisterns were built in 1872 and four more by 1875.[77] These provided water supply well distant

Guardian No. 2 steamer and hose cart in 1876. The fire company posed in their dress uniforms, likely for a parade or the ever-important annual fire department review. The city purchased the steamer from Clapp and Jones of Hudson, New York, in 1872. Behind it is a four-wheeled hose cart obtained from G. W. Hannis of Chicago for $700 in 1875.

Transition to Steam Fire Engines

from the rivers and were the only practical water sources for the steamers in those neighborhoods. A January 1875 review of "The City" in the *Green Bay Advocate* stated the cisterns "have proved effective in the extinguishment of fire," and that in these areas "previously if a fire occurred there was no alternative but to let the building burn."[78] A fire on Cherry Street illustrated the value of the cisterns when firefighters drew over 60,000 gallons of water from three interconnected cisterns.[79] Water supply on the edges of the city, away from the rivers, now matched the high volume water demand of the steamers.

Steps also were taken to improve water supply in the center of town. Two docks were built in 1874 over the Fox River at Doty and Cherry streets.[80] It is unknown if these docks were exclusively for fire department use or were also available for general use. Firefighters would park the steamers on the dock, drop the supply hose into the river, and then deploy attack hoses to the burning building. The Common Council paid to keep open holes in the ice at these river access locations during the winters.[81] Eventually, the fire department chief engineer was given the responsibility to arrange for keeping open holes in the ice.[82] Access to water through two-foot-thick ice on the Fox River was essential to saving a railroad roundhouse in the

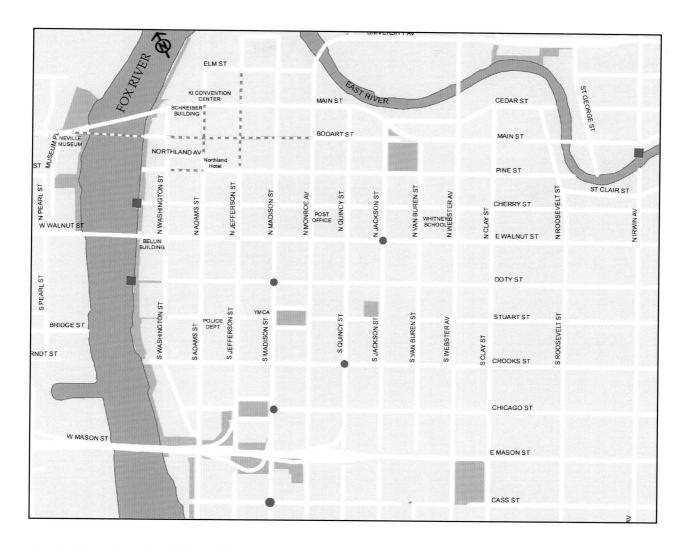

Fire department cistern and dock locations (1863-1875) within modern Green Bay. The substantial water volume demands of the two new steamers prompted construction of cisterns (circles) and docks (squares). Streets that no longer exist as of 2016 are indicated by dashed lines.

South Ward in 1881.[83] In addition, the city built a cistern near Engine House No. 2 on North Adams.[84] Between the cisterns and docks, GBFD had a substantial water supply system throughout the city.

As with the Enterprise No. 1 steamer, the size and weight of the Guardian No. 2 steamer made it impractical

Astor No.1 Engine House early 1900s. This engine house was built around 1852 on South Washington, just south of Walnut. To make way for a new, larger station for the steamer, the city moved this engine house in 1868 to this location at the southern end of the triangular park formed by the intersections of South Washington and South Adams. The squared-off base of the park and barn-type doors for the hand pumper fire engine are plainly seen. This view faces north up South Adams. This site is currently the lawn adjacent to the concrete apron of modern Green Bay Metro Fire Department Station No. 1.

for firefighters to haul the fire engine by hand. Therefore, the same horse team bounty system was implemented for Guardian No. 2 steamer: $10 for the first team that hauled the steamer and $5 for the team that hauled the hose cart.[85] However, at a special meeting of Guardian No. 2, members sought "to provide some means to prevent the engine from being left without a team to haul it to fires at night."[86] It seems horse team owners were reluctant to interrupt their sleep when the fire alarm bells sounded at night. Consequently, Guardian No. 2 purchased a horse team and kept in a barn near Engine House No. 2 from 7 pm to 7 am. Individuals were assigned to sleep at the station and served as horsemen in the event of a fire alarm. Guardian No. 2 fire company itself, rather than the city, also purchased beds for the drivers staying overnight.[87]

The horse team engaged in other money-making work during the day. One volunteer described how the horses would haul buses carrying travelers between train stations and hotels.[88] It appears the fire company kept the money made from this work as well as the bounty for hauling the steamer.[89] For unknown reason, the horses and harnesses were summarily sold in early 1877, not quite two years later.[90] GBFD relied upon the bounty system from then on. Though expensive, this remained the best option. One annual report stated the department spent $1,207 on horse team bounties.[91]

It seems certain teams were dependable and used very often.[92] One firefighter, William Ritchie, received a commendation for frequently, and promptly, providing his horse team.[93] This arrangement proved its effectiveness in October 1876. Enterprise No. 1 steamer sprayed water on the fire four and one-half minutes after sounding of

Transition to Steam Fire Engines

the alarm bell—an impressive time even by modern standards.[94]

However, delays still occurred. In 1875, Guardian No. 2 steamer was first to arrive at a fire that was much closer to Engine House No. 1. Enterprise No. 1 steamer apparently was delayed because "the heavy team generally used was over at the depot."[95] At another fire in October 1878, it took at least a half hour for either steamer to arrive, both having been delayed while waiting for horse teams to respond.[96] Similarly, a house only two blocks from Engine House No. 1 was "virtually destroyed" because nearly all the horses were gone from the livery (stable) near that station.[97]

The firefighters occasionally became inventive in their efforts to secure horses. In a 1936 interview, one former volunteer, August Delwiche, stated that in some circumstances, "the engine was usually hitched to any horses that happened to be nearby. The owner of the horses very often was left bewildered, because they [firefighters] did not make any explanation when they took the horses." The confused horse owners were later given the bounty.[98]

A unique problem with the bounty system cropped up in 1883 when two horses provided by Frank Hagen were burned at a fire.[99] Hagen eventually was compensated for his losses, but the process was difficult.[100]

Competition between the two steamer fire companies to be "first to the fire" was as real then as it is today. A broom displayed on the steamer or its engine

Franklin No. 3 fire company and hand pumper about 1871. The firefighters are in their formal dress, probably for a parade or inspection. The used hand pumper was purchased in 1871, though the manufacturer is unknown. Their helmets are adorned with the number "3" and several officers are carrying speaking trumpets, where were used to improve communication of commands.

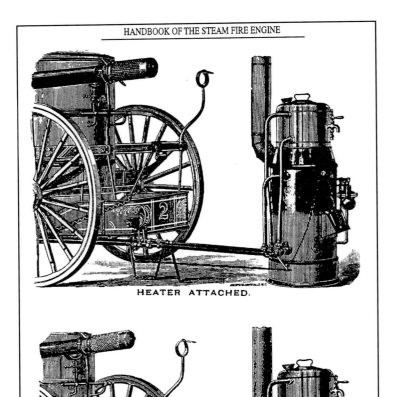

Schematic of a water heater and steamer fire engine. At the engine house, hot water from the apparatus floor water heater continuously circulated through the fire engine boiler. In the event of an alarm, firefighters quickly disconnected the water lines and started a fire in the steamer burn box. This process shortened the "time to steam." GBFD had water heaters in both steamer fire engine houses.

house signified they were the first to spray water on the fire.[101] After the fire alarm sounded, the firefighters would have been keenly aware the competition had started. Not knowing if or when horse teams would arrive must have been aggravating. The Common Council again considered purchasing horses dedicated to the fire department in 1881, but again took no action.[102] It was not until GBFD became a full-time department in 1892 that the city purchased dedicated horses.

Other issues regarding the steamers also were addressed. The water pumps on these fire engines were powered by steam produced by the onboard boiler. The ability to rapidly produce steam from the onboard boiler tank was necessary to flow water as soon as possible. An innovative solution reduced "the time to steam." When parked at the engine houses, the steamers were connected to stationary water heaters located on the apparatus floors. Hot water constantly circulated from the apparatus bay water heater to the steamer boiler through detachable water pipes. This kept the water in the onboard boiler tank of the steamer

86   Chapter 4

warm or even hot. When the fire alarm sounded, the engineer would immediately use paraffin, or oil-soaked rags and kindling, to light a fire within the steamer burn box. Once this fire developed, the engineer added coal. Thus, the already-hot water in the onboard boiler was further heated to boil, producing steam. By the time the steamer arrived at the burning building, steam pressure was fully developed to immediately generate water streams for firefighting.

Shortly after the purchase of Guardian No. 2 steamer, the city provided stationary, apparatus floor water heaters for both engine houses.[103] The utility of these heaters became obvious one summer when the heaters were not in use "as a matter of economy." Both steamers took longer than normal to "get to steam" from cold water at a major fire, creating a substantial delay in their ability to spray water.[104]

Having ample coal supply for the onboard boiler fire was another issue. A small bin on the back of the steamer provided an initial coal source. More was almost always needed, especially for long-duration fires. A small wagon was purchased which carried enough coal to fuel the steamers at the fire scene.[105] As with the steamers and hose carts, a bounty was paid for a horse team to pull the coal wagon to fires.[106] Though seemingly simple, this was an important function to support the steamer.

Through the end of 1875, transition to steamer fire engines had prompted dramatic changes to GBFD. The massive manpower requirements of the hand pumpers had been revealed as a critical limitation in November 1863, at what was then the worst fire in Green Bay history. Several subsequent disastrous fires occurred, finally prompting the Common Council to spend the significant sums to purchase a first steamer in 1868. Success of the first steamer directly led to a second purchase in 1872.

The superior capabilities of the steamer brought important changes. A new engine house was built for Enterprise No. 1 steamer. Both steamers were manned by full-time engineers, marking the first full-time, paid firefighters in Green Bay. Several large, underground cisterns were built to address the water volume demands of the steamers and expansion of the city away from the Fox River. This brief, but intense period of change occurred when the limits of the hand pumpers crossed with the technological innovation of the steam fire engine. GBFD had profoundly changed, emerging from simple to become state-of-the-art.

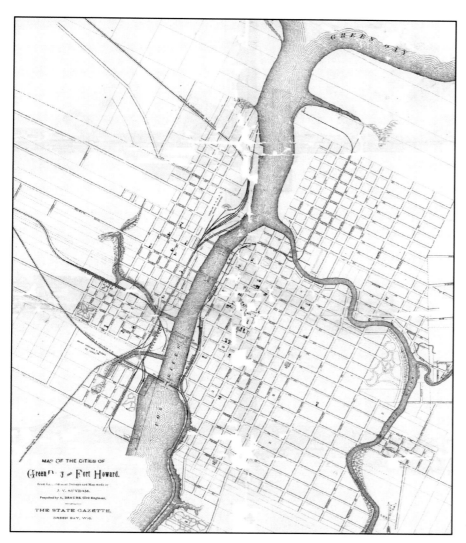

Map of Green Bay and Fort Howard from 1874. Green Bay in particular had spread further away from the Fox River, the primary source of firefighting water. This induced construction of numerous underground cisterns, specifically to store water for firefighting.

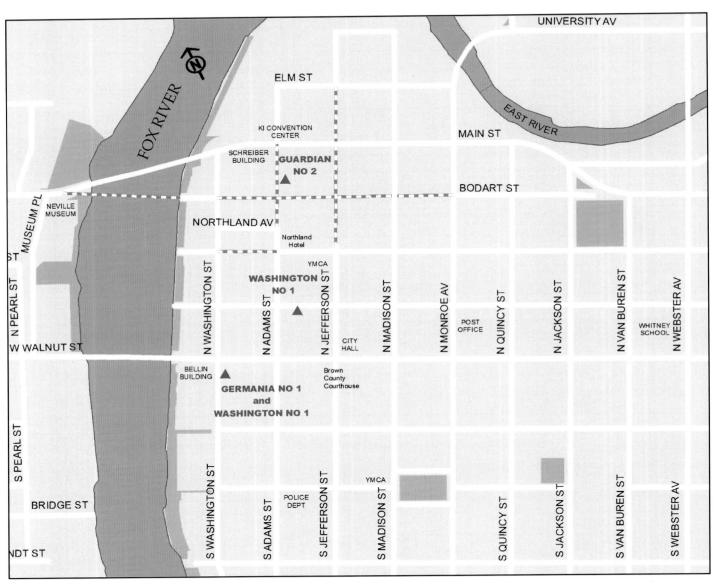

GBFD fire stations locations (1876-1886) within modern Green Bay. Fire stations discussed in this chapter are indicated by the triangles and labeled with the company names. Streets that no longer exist as of 2016 are indicated by dashed lines.

Chapter 5

Consolidation and Progress
1875 - 1886

The Green Bay Fire Department (GBFD) experienced a whirlwind of change after purchasing the steamers in 1868 and 1872. As a consequence, GBFD transformed from a simple frontier fire department to a modern one by 1875. A series of more moderate alterations took place over the course of the next decade, most of which were extensions of the just-realized successes.

The most significant change during this period involved the engine companies. At the start of 1875, there were four GBFD engine companies: Germania No. 1 and Guardian No. 2 with steamers, and Astor No. 1 and Franklin No. 3 with hand pumpers. Shortly after the steamers became operational, it was clear they far out-performed the hand pumpers. Not only did the steamers provide tre-

> **CITY NOTES.**
> —The Astor Fire Company No. 1 and Franklin No. 3 have been disbanded.

City Notes column from the August 21, 1875, *Daily State Gazette*. The two steamer fire engines (purchased in 1868 and 1872) far outperformed the hand pumpers of Astor No. 1 and Franklin No. 3. Now considered obsolete, these two companies were disbanded and the equipment sold.

"Where there are steamers it is next to impossible to get people to labor on the hand engines."

- *Green Bay Advocate*

mendous water flow volume, reach, and pressure, but they required fewer firefighters to operate. In contrast, the hand pumpers required many more firefighters to accomplish less. They had become obsolete.

It was difficult to justify financial support of the hand pumper companies in addition to the steamers. The Common Council first considered this issue at the same meeting they approved the purchase of Guardian No. 2's steamer.[1] The council took no action at this point, and the hand pumper companies remained active.

Subsequently, steamers and hand pumpers operated at the same fires. In these circumstance, the limitations of the hand pumpers were even more evident, especially with regard to the massive manpower requirements. One newspaper noted, "Where there are steamers it is next to impossible to get people to labor on the hand engines."[2] Thus, because of operational difficulties and the financial realities, the Common Council summarily disbanded Astor No. 1 and Franklin No. 3 hand pumper fire companies in August 1875.[3]

Although their fire companies were gone, the former members of Astor No. 1 and Franklin No. 3 were not ignored. The Common Council gave diplomas to those firefighters who had served at least five years. All were encouraged to join the three remaining companies: Germania No. 1, Guardian No. 2, or Washington No. 1.[4] It is unknown how many firefighters accepted the offer.

One newspaper expressed moderate concern that Green Bay now "has now to rely on the two steamers in case of fire, with the assistance of Fort Howard engine."[5] In spite of some initial unease, reliance on the steamers was well-justified. Water flow for fire suppression was much better and required far fewer firefighters. The improved capabilities of the steamers were a force of change that ultimately resulted in a warranted reduction in the number of GBFD fire companies.

The former Astor No. 1 and Franklin No. 3 sites and equipment remained city property, and the former engine houses often were used as voting sites. Fire alarms at these locations were still used to summon the fire department for many years. Eventually, in 1882, the Astor No. 1 alarm bell and Franklin No. 3 alarm triangle were given to the schools.[6]

The hand pumpers were no longer needed upon disbandment, and three were soon sold. The Astor No. 1 hand pumper, purchased in 1851 and used by Germania No. 1

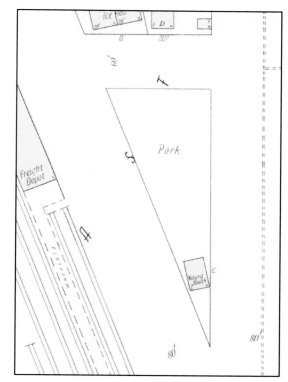

Former Astor No. 1 engine house serving as a "Voting Booth" from the 1900 Sanborn Insurance map. The engine house (built in 1852 and moved to this location in 1868) was at the southern end of the triangular park formed by South Washington (left) and South Adams (right). Modern Green Bay Metro Fire Department Station No. 1 is on the northern end of the triangular block.

Consolidation and Progress

and then Astor No. 1, went to the Village of Ahnapee (later renamed Algoma).[7] Today, this James Smith hand pumper is still with the Algoma Fire & Rescue Association. The Franklin hand pumper, purchased in 1870, was sold to the Village of Peshtigo.[8] Its ultimate fate is unknown.

GBFD had two other obsolete hand pumpers in addition to the two from the disbanded companies. Guardian No. 2 hand pumper was sold to the Village of Chilton.[9] The city obtained the hand pumper in 1858 and took it out of service after purchasing Guardian No. 2 steamer in 1872. Its fate also is unknown.

Old Croc, first purchased from the military in 1843, was not sold and never left Green Bay. It was used at a local saw mill for fire protection in 1878, and by 1891, it was stored behind Engine House No. 2.[10] Old Croc even made a brief appearance in a 1913 promotional movie about Green Bay.[11] Today, Old Croc, built by Harry Ludlum in New York in the early 1820s, is prominently displayed at the Neville Public Museum.

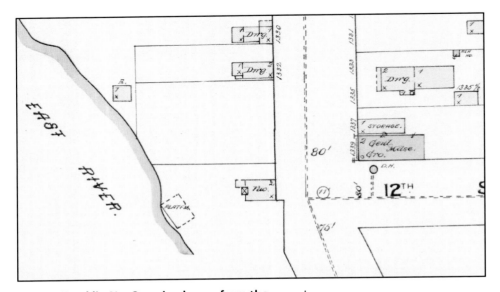

Former Franklin No. 3 engine house from the 1887 Sanborn Insurance map. Built in 1860 on Main Street at the intersection with 12th Street (modern North Irwin Avenue), it is shown here in yellow in the center of the image as vacant (Vac.). This building was demolished in 1887 to make way for a new station for the reformed Franklin No. 3 Hose Company.

Only a few changes occurred during the next several years after the hand pumper companies' disbandment in 1875. The city purchased a large, four-wheel hose cart, capable of carrying a great deal of hose, for Guardian No. 2.[12] Long hose lays of several hundred feet could now be deployed, fully utilizing the capabilities of the new Guardian No. 2 steamer. In 1876, the city purchased a new, purpose-built hook-and-ladder wagon for use by Washington No. 1 from G. W. Hannis Company of Chicago.[13] The old wagon, originally built in Green Bay, was sold to the neighboring village of De Pere.[14] Nothing is known about its history beyond that.

Firefighters had to address a hose maintenance issue. The outside jackets of the hoses often got wet at fires, and crews would attempt to dry them by hanging them within towers at the engine houses. The existing hose tower at Engine House No. 2 could not accommodate the 50-foot sections. The ends would lay on the floor, kinked, and could not adequately dry. The wet sections would then rot, ultimately leading to weak points. This manifested into a real problem at a September 1876 fire when seven sections of hose from Guardian No. 2 burst while on scene, all due to inadequate drying and rot.[15] To resolve this issue, the hose drying tower at Engine House No. 2 was made taller

Fire department announcement from the August 21, 1875, *Daily State Gazette*. Immediately after disbanding Astor No 1, Green Bay sold the Smith hand pumper to the village of Ahnapee, which today is the City of Algoma. The *Truesdell* traveled around the northern tip of Door County to reach its destination because the Sturgeon Bay canal did not yet exist.

Fire department news from the August 26, 1875, *Green Bay Advocate*. The paper recognizes the services of the two recently disbanded hand pumper companies, but in reality there was uncertainty if the two steamer fire engines were adequate.

Consolidation and Progress

James Smith manufactured hand pumper fire engine. This machine arrived in Green Bay in 1851 and was used by Alert, Germania No. 1, and Astor No. 1 GBFD fire companies before being sold to the newly formed fire department in the village of Ahnapee, Wisconsin, which later became the city of Algoma. The Luxemburg (Wisconsin) Fire Department used this hand pumper from 1908 to 1934. It was later returned to Algoma and restored.

so hoses would hang off the floor and dry completely.[16]

Another noteworthy change involved staffing. As GBFD developed in complexity and sophistication, more demands were placed on those in charge. Accordingly, the chief engineer (in modern terms, the fire chief) was made a part-time position in 1876, as was the assistant chief engineer two years later.[17] The chief engineer salary, initially $50 per year, was raised to $100 by 1884. The assistant chief always made half that amount.[18] An 1875 ordinance listed the duties and authorities of these officers in great detail.[19]

Each year the fire companies nominated the two chiefs, who were then confirmed by the Common Council. The nominations usually were unanimous amongst the fire companies. Subsequent Common Council action was just a matter of formality. However, there were a few years – 1878, 1879 and 1881 – when the fire companies did not come to a consensus and separate nominations were presented to the Common Council.[20] In these cases, the Common Council selected by ballot the two chief officers from the nominees. Chief engineer and assistant were maintained as part-time positions, nominated by the fire companies, until GBFD became a full-time, paid agency in 1892.

Placing the steamers into service resulted in fewer firefighters in the department. In 1874, two years after the second steamer arrived, there were 140 Green Bay firefighters.[21] There were 131 firefighters the next year, and by 1877, GBFD consisted of only sixty-seven volunteers.[22] This trend con-

Hose cart belonging to the Algoma Fire and Rescue Association in 2015. Green Bay sold a hose cart along with the Smith hand pumper to Ahnapee in 1875. There are no markings confirming that this is the same hose cart. However, the style suggests this may be the former GBFD Astor No 1 hose cart. Firefighters hauled the hose cart by hand using ropes from the reels under the frame. Others steered using the arm extending forward. Fire hose was deployed from the large reel between the two wheels. A tool box is in the back.

Consolidation and Progress

tinued, and by 1880, the number of firefighters had plummeted to thirty-seven, including the two full-time steamer engineers and the two part-time chiefs.[23]

Old Croc hand pumper at the Neville Public Museum in the 1930s or 1940s. First purchased from the US military in 1842, this hand pumper fire engine served with several GBFD companies. The city loaned it to local businesses when hand pumpers became obsolete. This first GBFD fire engine is nearly 200 years old and is still part of the Neville Public Museum collection in Green Bay.

Decreased GBFD membership was not surprising given only a few firefighters were needed to operate the steamers, while the hand pumpers required scores to work the brakes. The disbandment of the two hand pumper fire companies in 1875 was another factor.

Municipal rules even reflected the reduction in GBFD membership. Specifically, the renewed City Charter and Ordinances, passed in 1882, limited each fire company to fifteen members.[24] Consistent with this standard, an insurance report in 1883 recorded forty-five volunteers among the three GBFD companies.[25]

Unfortunately, the loss of firefighters continued. In early 1884, following a large, difficult downtown fire, the chief engineer reported to the Common Council that GBFD needed more firefighters.[26] His account stated there were now only thirty members between the one hook-and-ladder and two steamer companies, substantially less than the year before. Regrettably, the council did not heed his suggestion.[27]

From 1885 to 1887, GBFD staffing levels continued to hover around thirty volunteers, plus the four full-time and part-time staff.[28] A slight increase to about forty firefighters occurred when a new hose company formed in 1887.[29] However, controversy surrounding poor performance by the entire fire department at a large fire caused volunteer GBFD membership to plummet to twenty-five by 1892.[30]

The Common Council provided direct financial support to the fire companies for the first time in 1878. The city previously had paid bills for equipment, apparatus, services, and stations, but never provided money straight into fire company funds. With decreasing membership, the burden on the remaining volunteer firefighters prompted the fire companies to ask for funding.[31] This money would serve to compensate the firefighters, hopefully curbing the loss of its current members and aiding in recruiting even more. Arguing that "a volunteer fire department in this city is for certain reasons much preferable to a paid one,"

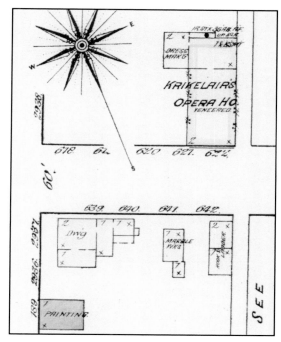

Washington No. 1 house from the 1883 Sanborn Insurance map. "The "Hooks" house was built in 1860 on the south side of Cherry Street in the middle of the block between Adams (left) and Jefferson (beyond image to right) and closed in 1883. It is depicted here in the center right of the image, adjacent to an alley and across the street to the south of Krikelair's Opera House.

Consolidation and Progress

> **Fireman's Party.**—Guardian Eng. Co. No. 2 will give a party at Klaus' Hall, on Friday evening, Dec. 18, at 8 o'clock. Firemen of Green Bay and Fort Howard, and the public generally, are respectfully invited.—Tickets $1.00; to be had at *Advocate* office and M. Rusch's. *Per Order Committee.*
> Dec. 9. 1868. 20-2w

Party notice for Guardian No. 2 from the December 17, 1868, *Green Bay Advocate*. Parties and dances served as a fundraiser for the volunteer fire companies. Many events were held at Klaus Hall, on the south side of Pine Street between North Washington and North Adams. Urban renewal eliminated that block in the 1970s. Fire companies used Klaus Hall so often it was also known as Firemen's Hall.

the Common Council gave $45 to each of the three companies.[32] This practice continued until GBFD became a full-time, paid fire department in 1892.[33]

Another new source of fire department funding came from fire insurance premiums. Wisconsin law required insurance companies to donate two percent of fire insurance premiums to local fire departments.[34] The first instance of a payout in Green Bay was in 1873, when $557.70 was divided between the five fire companies.[35] The amount from insurance premiums fluctuated the first few years from $743 in 1875 to $369 in 1879.[36] Through the 1880s, insurance payouts to the fire department rose steadily to between $600 and $700.[37] The last disbursement of $915.89 to the volunteer department came in 1891, shortly before conversion to a full-time, paid department.[38] A portion of these funds likely was distributed as firefighter compensation.

A surprising source of funding came from the volunteer firefighters themselves. Surviving records from three volunteer fire companies reveal the firefighters routinely paid dues and fines. Records for Guardian No. 2 during the hand pumper era (1859-1863) still exist and show such mandatory dues.[39]

Washington Hook and Ladder Company No. 1 in 1881. This hook-and-ladder wagon (purchased for less than $1,000 in 1876) features a tiller driver—the firefighter at the rear operated a steering wheel controlling the rear axle. Normally the wagon was pulled by firefighters, but it is horse-drawn in this image. This likely was on the occasion of a parade or annual inspection, which is why the firefighters are in formal dress uniforms.

Even more records exist from the time after the steamers arrived. There are numerous examples of fines and dues imposed on the firefighters. New members of Guardian No. 2 paid a 50-cent initiation fee. All firefighters paid 10 cents at each Guardian No. 2 meeting.[40] When Franklin No. 3 was reformed in 1887, 40 cents was charged as a special assessment at the first meeting. Those joining later paid 25 cents.[41]

Money also was raised from fines levied on firefighters. For example, missing a monthly meeting cost 25 cents for Guardian No. 2, 10 cents for Washington No. 1, and 10 cents for Franklin No. 3.[42] Fines for missing alarms,

Consolidation and Progress

whether for a false or real fire, cost the offender 25 to 50 cents.[43] Some other fines imposed included 25 cents for failure to clean the engine house when assigned and $2.50 for missing the always-important annual city inspection day.[44] Washington No. 1 imposed $1 fine for missing truck-cleaning duty.[45] While these fines raised money for expenses, they also served to promote attendance at fire company events.

The amount of money raised from dues and fines was substantial for the time. In the mid-1870s, the treasurer for Guardian No. 2 would collect anywhere from $2

Uncle Frank's Block in 1873 looking north up Washington Street from Cherry. The fire limits ordinances applied to the congested downtown and mandated only brick or masonry exteriors. All of the two- and three-story buildings in this image comply. However, there are a couple of wooden-exterior, one-story buildings. These likely were moved to this block between its complete destruction in the November 1863 conflagration and passage of an 1864 ordinance prohibiting such practices.

to $30 at each meeting.[46] In the early 1890s, the Franklin No. 3 treasurer collected $1 to $9 per meeting.[47] Washington No. 1 charged $5 to $36 per member per year, with an average of about $25.[48]

Accumulation of excessive, unpaid fines and dues resulted in expulsion from the fire company.[49] Franklin No. 3 set $1 as the amount of accumulated debt that resulted in dismissal.[50] In at least one instance, the newspaper announced on behalf of Guardian No. 2 that members with delinquent accounts were to settle or be stricken from the company roll.[51]

One more fundraising mechanism, the fundraising dance, started in 1856 and continued long after conversion to steamers. Every fire company held dances, with the proceeds going to company funds. Guardian No. 2 raised $100 at an 1876 dance and $38 two years later.[52] An 1887 dance raised $123 for newly re-established Franklin No. 3.[53] Dances for Washington No. 1 raised $36 and $13 in 1889 and 1890, respectively.[54] In at least two instance in the mid-1880s, large dances were held on behalf of the entire GBFD following the annual fire department inspection. The 1883 dance netted $115.[55] Notices for these parties appeared in the newspapers. In the case of a Guardian No. 2 dance in 1878, the *Green Bay Advocate* stated these

> **Firemen's Dance.**—The Hose boys of Guardian Engine Co., No. 2, are to have a dance on Friday evening of this week.

Dance notice for Guardian No. 2 from the December 16, 1869, *Green Bay Advocate*.
It seems the fire company was divided into hose and engine groups. In this case, the "Hose boys" were hosting the party.

Consolidation and Progress

The 100 block of South Washington, with the Walnut Street Bridge beyond, in 1891. Most buildings have only brick or masonry exteriors in accordance with the fire limits ordinance. Several wooden-exterior structures along the riverbank were built before the ordinance passed. Engine House No. 1 is to the right with the bell tower.

parties "offer the only avenue through which we show any return of gratitude for the services which the Green Bay firemen so generously render through the year."[56]

Two water supply issues were addressed shortly after the steamers arrived in 1868 and 1872. The steamers required large volumes of water and the city had expanded away from the unlimited resources of the river, particularly to the east. By 1875, the city had built at least nine underground water tanks, also known as cisterns, and three fire department river docks.

However, because neighborhoods had developed away from existing cisterns and docks, even more were needed. Four more cisterns were built by 1884, the last of which was located at Day Street and modern North Irwin Avenue, at what was then the northeast corner of Green Bay.[57] One former volunteer firefighter recalled that the cisterns were known by the nearest residence, such as Howe, Bishops, Van Dyke, Lenz, Erdman, or Nick "tanks" rather the street location or intersection.[58] More fire department docks were built on the Fox and East rivers.[59] One was at the foot of Eliza Street on the south side of the city, another at Webster and the East River to the north. Three were built downtown to provide water supply points in the commercial area. The cisterns and docks proved adequate for these areas.

The former Cook's Hotel in the 1920s, known at this time as the Sherwood Hotel. This hotel, built in 1875 at the southwest corner of North Washington and Cherry (the current site of Nicolet National Bank), featured heat detectors in each room, which rang a bell at the front desk when activated. The builders installed these devices even though not required by law at the time.

Consolidation and Progress

Station No. 1 on South Washington Street in 1902. The section to the right was built in 1868 to accommodate the Enterprise steamer. Washington No.1 used the 1883 addition to the left. City government used the upstairs rooms of the addition for offices and meetings. This station remained in service until 1929. Today, this site is the southern end of the Backstage at the Meyer. The building behind the utility pole is the site of the Meyer Theatre.

Some lesser-developed areas of the city still remained too far from the rivers and cisterns. For example, in early 1884, GBFD started to respond to a fire on the north side of the East River, at what was then the remote edge of the city. The firefighters never made it to the fire, though, due to the fact the home was "so far distant and no water to be had there."[60] The next month, in the same neighborhood, a home burned and GBFD was not even alerted. Distance and lack of water supply made a response pointless.[61] The same thing occurred on the east side of the East River in late 1885.[62] In spite of some unprotected areas, the cisterns and docks were the best option available until a waterworks system with underground mains and hydrants was constructed.

Not only had GBFD's suppression operations improved, but the fire limits ordinances changed for the better as well. First established in 1859, this municipal law essentially prohibited wooden exteriors in the congested downtown. All building exteriors had to be stone, masonry, or metal so as to inhibit the spread of fire from building to building in the crowded business and mercantile area. The city updated the fire limits ordinance in 1864, 1867, and 1872, mostly to include expanded areas of downtown.[63] One interesting loophole was addressed: moving wooden-exterior buildings into the fire limits area became expressly prohibited.[64] Another update required property owners and occupants to keep outdoor areas clear of flammable materials.[65]

The value of these initial fire limits rules was demonstrated in 1872, when a wooden shed in the fire limits area caught fire, but the flames did not extend to the nearby brick-sided buildings on Washington Street.[66] Similarly, a store at the congested northwest corner of Pine and Adams was destroyed by fire. However, there was no fire extension to the abutting buildings because of fire-resistant exteriors. The fire was limited to the building of origin.[67]

Recognizing the success of these codes, the fire limits ordinance was substantially updated in 1875.[68] The most significant changes were new, detailed construction requirements. Other issues addressed included the first

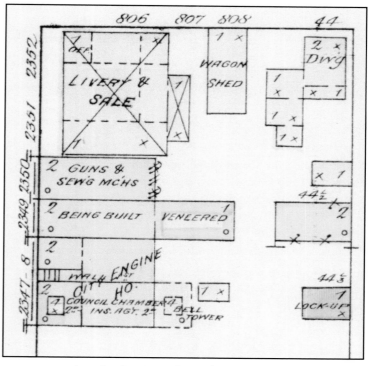

Station No. 1 from the 1883 Sanborn Insurance map. The station is in the lower left, facing South Washington to the left. Walnut is at the top of the image.

Consolidation and Progress

exit rules, storage of kerosene, and use of spark arrestors on the exhausts of coal-burning machines.[69] Another new provision required bonfires to be at least 100 feet from any building. Because of the frequent and large fires at lumber yards, these businesses were now prohibited within the fire limits area.[70] All of these rules were reaffirmed in 1882, and ordinances added even more detail to construction requirements.[71] Fire prevention laws were well-established and in-depth.

Though not mandated by law, one unique change in Green Bay reflected the growing concern for fire safety. Cook's Hotel opened in 1875 on the southwest corner of North Washington and Cherry (site of present-day Nicolet Bank). Each room in the four-story building had a heat detector that triggered at 120 degrees. It sent an electrical signal that rang a bell in the hotel office and alerted the staff, who would then summon the fire department. This is the first recorded instance of an alarm system in Green Bay.[72]

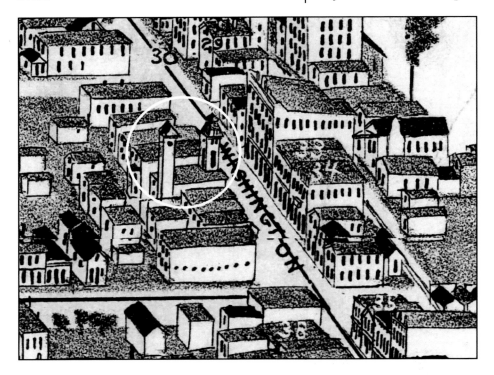

Station No. 1 on South Washington Street from the 1893 lithograph. This view faces south with Walnut in the foreground. The fire station had a bell tower in front and a hose drying tower in the rear.

108   Chapter 5

By the early 1880s, GBFD consisted of three fire stations: Germania No.1 house on South Washington, built in 1869; Washington Hook and Ladder No. 1 house on Cherry, built in 1860; and Guardian No. 2 house on North Adams, built in 1859. At a May 1881 Common Council meeting, all three stations were reported to be in need of repair. Guardian No. 2 house was "in very bad condition" and was extensively rebuilt in 1883.[73] At the same time, a large addition to Germania No. 1 house was built, essentially doubling the size of the station. Washington No. 1 Hook and Ladder Company left its former station and now used the new apparatus bay of the addition. Part of the upstairs was used for fire department meetings.[74] With Washington No. 1 now at the Germania No. 1 South Washington station and Guardian No. 2 house still on North Adams, Green Bay had only two fire stations. Both stations were provided with telephones, a recent innovation.[75]

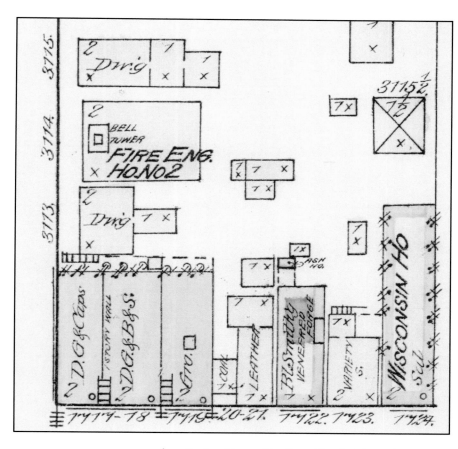

Engine House No. 2 on North Adams from 1883 Sanborn Insurance map. This engine house (left-center) was originally built in 1859. It was comprehensively rebuilt in 1883, and then served as a volunteer GBFD station until 1892. It faces North Adams (left), one-half block south of modern Main Street (then Cedar), beyond the top of this image. This section of North Adams was eliminated in the 1970s.

Consolidation and Progress

Some of the second-story rooms of the Engine House No. 1 addition were designated for other city use. Previously, municipal officials and the Common Council had used rented rooms to perform their functions. Now the city government had dedicated offices and meeting rooms, albeit in the upstairs of a fire station.[76]

Members of Germania No. 1 in 1882. The caps have the fire company logo "Rough and Ready." Standing left to right are John Kittner, Conrad Silberdorf, Ed Fuerst, and Charles Pfotenhauer. Sitting left to right are John Bertles, Ed Kittner, and Henry Loewert. Sitting on the floor is Herman Scheller. Bertles was Enterprise No. 1 Engineer of the Steamer from 1872 to 1885, John Kittner was his assistant, and Ed Kittner was GBFD Chief Engineer (in overall charge of the department, modern term is fire chief). Ed Kittner operated a wagon repair shop on South Washington, the current site of Kittner's Bar.

Obtaining the two steamers in 1868 and 1872 initiated tremendous change for GBFD through 1875. In comparison, the next decade brought about a series of gradual, almost mundane modifications. Most significantly, the two hand pumper fire companies disbanded and the fire engines were sold. The chief engineer and assistant became part-time positions. Volunteer membership declined, mostly due to reduced need for manpower. However, the number of firefighters became a concern, and financial compensation was offered to attract and retain. An addition to Engine House No. 1 made room for the hook-and-ladder company, leading to closing of the

Intersection of North Adams and the former Main Street, March 1881. Guardian No. 2 engine house is on the left with the bell tower. This section of Main Street, going left to right, was eliminated in the 1970s. Modern Main (at that time Cedar) would be about one-half block beyond the engine house. In the close-up at right, the number "2" can be seen on the facing of the engine house above the windows, signifying the fire company number.

Consolidation and Progress

original ladder house on Cherry Street. Combined with the closing of Astor No. 1 and Franklin No. 3, GBFD went from five stations in 1875 to only two in 1883.

Although seemingly a reduction in force, in reality, GBFD became streamlined, sophisticated, and modern. It barely resembled itself from just twelve years before. However, in 1880, GBFD would face the greatest fire ever in the city—a devastating firestorm. This conflagration would challenge the improved GBFD like no other fire before – or since.

Fire department cistern and dock locations (1875-1886) within modern Green Bay. The substantial water volume demands of the two new steamers prompted construction of the cisterns (circles) and docks (squares). These are in addition to the cisterns and docks constructed earlier, shown in the map on page 81. Streets that no longer exist as of 2016 are indicated by dashed lines.

Consolidation and Progress

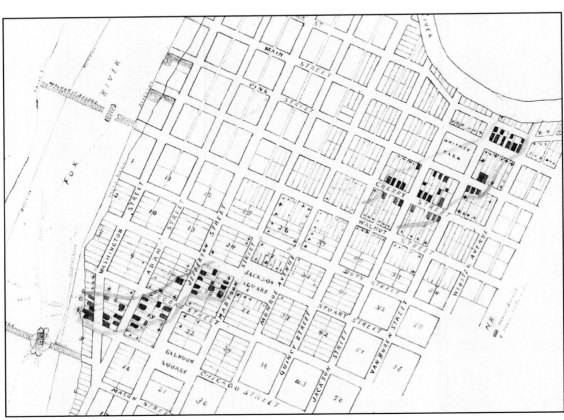

Survey map prepared by J. V. Suydam for an 1882 lawsuit following the Great Fire of 1880. This image shows Green Bay with the Fox River on the left, East River in the upper right, along with the Mason Street Bridge on the bottom left and Walnut Street Bridge in the upper left. The solid boxes are burned buildings and the "+" indicates buildings threatened by the fire, but survived. The highlight outlines the damaged areas and shows the wind direction.

Chapter 6

The Great Fire of 1880

In the 1800s, it was not uncommon for American urban areas to suffer conflagrations — massive fires that moved from building to building, block to block, leaving great swaths of devastation. The most widely known conflagration remains the Great Chicago Fire of 1871. That same day in Northeast Wisconsin, the entire village of Peshtigo was annihilated by fire, along with significant areas across the bay northeast of Green Bay and up the Door peninsula. Other, lesser-known Wisconsin conflagrations occurred in urban areas during the late 1800s in Oshkosh, Racine, Marshfield, and Milwaukee, all of which resulted from a catastrophic combination of combustible construction materials, dry, windy weather, and overwhelmed firefighting efforts.

Headline from the September 23, 1880, *Green Bay Advocate*.

> *"You could see only fire, nothing of the boat."*
>
> - Resident who witnessed the *Oconto* going upriver

These same conditions led to a conflagration in Green Bay on September 20, 1880.

The fall of 1880 had been generally dry. There had been no rain for at least a week prior to September 20.[1] On that fateful day, a steady wind blowing from the southwest was so strong that it was "difficult to walk against" and was later estimated to be 25-30 miles per hour.[2] The coal-powered, propeller-driven ship *Oconto* was docked on the east bank of the Fox River at the foot of Pine Street. It pulled away from the dock en route to De Pere. However, heading into the strong wind made maneuvering into the channel difficult, and it was clear to witnesses on shore that the *Oconto* was struggling.[3] The crew added more coal to the boiler firebox in an effort to provide sufficient power for the ship's propeller. This resulted in a considerable amount of burning coal cinders to be expelled with the exhaust from the stack. The burning cinders landed along the riverbank, causing several fires and even "one man's clothes caught fire on his back."[4]

As the *Oconto* slowly moved upriver, burning coal cinders carried by the strong winds continued to land beyond the shoreline. A witness described how he "felt hot cinders and very thick smoke on my face and arms," while nearby a man extinguished several small fires.[5] Watching

The Oconto. On September 20, 1880, this boat attempted to travel upriver to De Pere against a very strong wind. Burning coal embers in the exhaust landed on the riverbank and started The Great Fire of 1880 in Green Bay.

the *Oconto* continue to labor against the wind farther upriver, one witness described, "Sparks of fire coming out of the smoke stack."[6] Even a deckhand on the *Oconto* testified at a later legal proceeding that the top of the smokestack appeared on fire, with sparks as big as coins landing on the deck and actually burning a hole in his clothes.[7] A resident on South Washington Street stated, "You could see only fire, nothing of the boat."[8] The *Oconto* left a trail of coal cinders that were "plainly seen" reaching the shore.[9]

Advertisement for the Astor Planing Mill from the 1874 Green Bay *City Directory*. The Great Fire of 1880 started in a wood shaving pile behind this unoccupied, three-story mill at the northwest corner of South Washington and Chicago.

As the boat passed farther upriver through Green Bay, this volley of fire continued to descend on the riverbank along South Washington. The *Oconto* strained past the Walnut Street Bridge. A woman living on South Washington near Crooks Street saw sparks strike the roof of her home, igniting several small fires that she successfully extinguished.[10] A nearby woodshed also caught fire, but this, too, was stopped.[11] The *Oconto* continued upriver, as did the barrage of burning coal cinders upon the buildings along South Washington. The captain and crew were seemingly unaware or indifferent.

Eventually, a sustained fire took hold at the Astor Planing Mill, located on the riverbank of South Washington at the northwest corner of the intersection with Chicago Street.[12] This three-story building had been out of operation for at least three months and therefore was unoccupied on September 20.[13] A slab dock at the mill extended from the shoreline, where a man was fishing. The *Oconto* passed by and he saw it "laboring heavy" and "throwing out coals and sparks" as large as coins that even reached South Washington, 500 feet away.[14] Estimates made for a subsequent lawsuit calculated the *Oconto* came as close as 300 feet from the planing mill dock.[15]

Burning coal cinders landed on a 110-foot-long pile

of highly combustible wood shavings covering the mill dock and extending to the mill structure.[16] A bridge tender on the Mason Street Bridge saw that after the *Oconto* had passed, the shaving pile in the rear of the mill was on fire.[17] Several others nearby made the same observation.[18] Upon seeing this moderately small fire in the shavings pile, one man ran three blocks north to Engine House No. 1, on South Washington near Walnut, to alert the Green Bay Fire Department (GBFD).

At about 2:30 pm, within a few minutes of the fire starting in the shaving pile behind the mill, the fire alarm bell on top of Engine House No. 1 rang.[19] Soon afterward, the fire alarm also sounded across the river, activating the Fort Howard Fire Department.[20] Firefighters were responding while the fire was still in its early stages. It was still possible to stop the fire, but a water supply problem soon would thwart even their best efforts.

Drawing of the fire extending from the Astor Planing Mill from the March 5, 1960, *Green Bay Press-Gazette*. This represents the early stage of The Great Fire as the blaze extends across South Washington. It erroneously shows the mill adjacent to the Mason Street Bridge. The mill was actually one block north.

First Presbyterian Church destroyed in in the Great Fire of 1880. Built in 1838, it was located on the east side of South Adams, a few lots south of Crooks.

The first fire company on scene was Enterprise Steamer No. 1. Assistant Foreman John Kittner reported the steamer stopped at a cistern at South Washington and Crooks, one block from the planing mill.[21] Fed by an underground pipe from the Fox River (about 150 feet away), this cistern was essentially an underground water tank that could hold up to 7,000 gallons.[22] In 1880, Green Bay did not yet have water supply from hydrants. The rivers and underground tanks (cisterns) were the only possible water sources. Firefighters dropped the steamer intake hose through a small access hole into the cistern and deployed an attack hose line down South Washington.

Tragically, the cistern was already nearly empty, resulting in a weak water stream from the attack hose that lasted only a few minutes.[23] The firefighters were understandably astonished. Because of the dry weather and strong wind pushing the water out of the Fox River and into the bay, the river level was estimated to be three to five feet lower than normal.[24] Engineer J. F. Bertles, who was operating the steamer, testified the water "was extraordinarily low," and as a result, "it didn't run through the supply pipe into the cistern."[25] Because the river level was below the intake pipe opening, no water could enter the cistern and any amount previously there had already flowed outward into the Fox River.

The firefighters repositioned the steamer on a dock, dropping the intake hose directly into the Fox River.[26] However, getting water from the dock to the steamer proved essentially impossible because the river water level was so low. In fact, the water was so shallow that firefighters had to place the supply hose in the mud. A firefighter worked in the weeds and muck, but his efforts to keep the supply hose from clogging were futile.[27]

The empty cistern and shallow river level at the dock crippled firefighting efforts during the critical initial stage of the fire. Bertles stated, "The engine couldn't do effective work around there at any time on account of these delays."[28] Kittner watched, while struggling at the cistern,

> *"The engine couldn't do effective work around there at any time on account of these delays."*
>
> - Enterprise Steamer No. 1
> Engineer J. F. Bertles

> *"If the water hadn't given out and these delays hadn't happened, the fire would have been extinguished at this time."*
>
> - Firefighter Louis Mohr

as flames from the shaving pile extended to the planing mill structure itself.[29] Strong winds fanned the blaze and firefighters looked on as the "flames rolled from the Astor Planing Mill".[30] Firefighter Louis Mohr, waiting at the useless nozzle on South Washington, watched in frustration as the fire extended from the planing mill. Mohr testified, "If the water hadn't given out and these delays hadn't happened, the fire would have been extinguished at this time."[31] A newspaper report two days later observed the "firemen were bravely but ineffectively fighting the flames near their origin."[32] Once the mill became fully involved, the fire had a foothold. A conflagration was developing.

The fire grew quickly and massively during what ultimately became a failure to establish water supply to the steamer. Within five minutes of the flames entering the back of the mill, the wind pushed blowtorch-like flames out the front.[33] The home immediately to the north caught fire, and flames spread to the next two homes along the west side of South Washington. The fierce wind carried burning debris across South Washington, setting fire to a large icehouse, a large "old bank," and two more homes. A firestorm—a conflagration—had truly begun.

Burning debris thrown to the northeast by the strong southwest wind landed on nearby buildings that

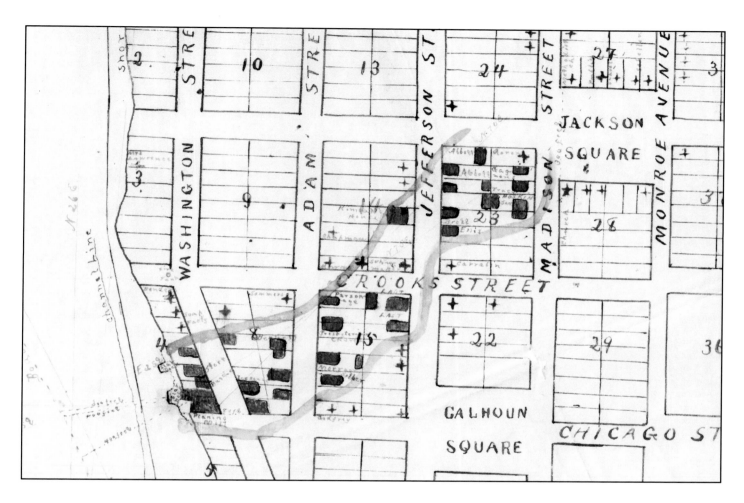

quickly became engulfed in flames. Three more dwellings on the South Adams Street side of the block caught fire. The fire then extended to the east side of Adams, where two houses, the First Presbyterian Church, and its adjacent parsonage burned. From there, firebrands thrown by the strong wind landed on nearby roofs and set seven houses afire on the block northeast of Crooks and Jefferson.

Close-up of the southern section of the conflagration area from the 1882 lawsuit map (page 114). The Astor Planing Mill, fire origin site, is at the most lower left. Firefighters attempted to use the empty cistern at the intersection of Washington and Crooks. To the upper right, Jackson Square and the four open lots to the northwest along Madison Street formed a sort of urban fire break that gave firefighters the opportunity to stop the spread of the fire.

When Elisha Morrow heard the alarms from downtown, he hurried to his home on the northeast corner of Crooks and Adams. He stayed on the roof of his house as "the air was nearly filled with cinders and sparks," and saved his home because he "kept the roof of my house wet until the danger passed."[34] Morrow's house survives to this day as the home of Captain's Walk Winery.

Firefighters used two steamer fire engines to make a stand against the firestorm at the intersection of Crooks and Adams.[35] They saved two houses, including the Morrow home, and stopped the fire from spreading north. But the firestorm continued its relentless advance to the northeast toward Jackson Square.[36] From his vantage point on the roof, Morrow saw the fire spread from house to house.[37] Later testimony established that between thirty and sixty minutes passed from the time the planing mill caught fire until the fire spread to the block just southwest of Jackson Square. Once there, eight homes and their outbuildings in that one block began burning simultaneously.[38]

Fortunately, the open square and many empty lots in the immediate path of the fire provided an opportunity to stop it. Essentially, these acted as urban firebreaks and the firefighters took advantage of this circumstance.[39] Steamer No. 1 was placed at a cistern on Madison Street.

ST. FRANCIS XAVIER CATHEDRAL, GREEN BAY.

This spring-fed cistern was full of water and supported significant firefighting efforts. Hose streams protected the Methodist Episcopal Church at the northwest corner of Jackson Square, as well as the Garon home at the southwest corner of Madison and Stuart, even as eight houses on the same block were consumed.[40]

Many residents assisted in the firefighting efforts.

St. Francis Xavier Cathedral in 1881. This church is located one block north of Jackson Square, where firefighters stopped the southern portion of the conflagration. The cathedral was in the path of the rain of burning debris thrown by the strong wind and survives to this day.

The Great Fire of 1880

The Elisha Morrow House (built in 1857) in 1884. Located at the northeast corner of South Adams and Crooks, it is still standing today as Captain's Walk Winery. Morrow threw buckets of water from the cupola (top-center of the house, behind the trees) to extinguish burning debris landing on the roof. Fire destroyed many homes nearby, including the First Presbyterian parsonage previously located in the foreground. The damaged lawn, scars on several trees, and the dead tree to the right are effects of the fire from four years earlier.

A firefighter reported that "every house more or less had people protecting it."[41] Using ladders and buckets, these homeowners kept the roofs wet and immediately doused burning debris that landed on the combustible shingles.

At this point, at least twenty-eight buildings, as well as some horse barns and other outbuildings, were burning between the river and Jackson Square.[42] The airborne storm of burning debris was incredible. Louis Scheller, protecting his home on Monroe Avenue, testified, "You might say it rained fire".[43] Another man, who defended his house at Quincy and Stuart stated, "The brands were very thick, the air was full of them," and some were carried higher than sixty-foot trees.[44]

The cistern on Madison provided water for over ninety minutes to Steamer No. 1, allowing firefighters to protect buildings in the path of the fire and prevent further progress of the conflagration.[45] The determined efforts of firefighters and Green Bay's highly motivated citizens successfully checked the advance of the fire at the intersection of Madison and Stuart on the west side of Jackson Square.[46] Unfortunately, the fire was not done in Green Bay.

A "tremendous gale" extended the blaze to the northeast.[47] About forty structures burned between the planing mill and Jackson Square, and the wind sent "showers of burning coals" many blocks distant.[48] Later testimony established the natural wind "heavy" at about 30 mph. However, accounts of the wind being "almost a tornado" suggest the fire itself contributed to the force of the wind, an effect seen many years later during the ae-

Next-day headline from the September 21, 1880, *Daily State Gazette*, a Green Bay newspaper.

rial fire bombings of urban centers during World War II.[49] The strong air currents generated by the fire added to that of the natural wind, creating the effects observed. As a result, the wind carried an immense amount of burning debris to the northeast, well beyond Jackson Square.

People actively protected their homes and businesses within the path of this rain of fire. One account described the "housetops for blocks in advance of the line of fire were dotted with men, throwing water upon roofs".[50] Homer Harder was so concerned he moved his belongings from his home two blocks from Jackson Square to the safety of a friend's home across the river in the Borough of Fort Howard. He returned to protect his home all afternoon, as it was "in a direct line of the sparks".[51] One resident observed that downwind of Jackson Square, "the air was full of burning cinders."[52] This torrent of firebrands in the air would have a tragic effect in other parts of Green Bay.

Along that direct line of burning debris was the block of Cherry Street between Quincy and Jackson streets. Here, many residents threw water to extinguish the burning debris landing on the roofs. However, no one was home to fight the storm of fire at 716 Cherry Street, the residence of Charles Kitchen.[53] Several neighbors busy protecting their own homes nearby watched as burning

> *"Housetops for blocks in advance of the line of fire were dotted with men, throwing water upon roofs."*
>
> *- Green Bay Globe*

debris landed on the roof of the Kitchen home.⁵⁴ As was common in that era, the roof likely was constructed with cedar shingles, which quickly caught fire.⁵⁵ The small fire grew swiftly, and within ten minutes the entire Kitchen house was engulfed.⁵⁶ One neighbor later lamented that with only a ladder and four or five pails of water, the fire on the Kitchen roof could have been stopped.⁵⁷ However, everyone nearby was fully occupied protecting their own homes.

It was later determined that the Kitchen home was 1,650 feet from the closest burning house in the southern section.⁵⁸ Also, there were fifty-seven buildings that did not catch fire between the last burned house near Jackson Square and the Kitchen home.⁵⁹

From the Kitchen home, the flames spread and destruction began anew. The blaze generated a second, northern section of the firestorm that spread with "great rapidity" from building to building, block to block, causing even greater devastation than what had occurred in the southern section.⁶⁰

The fire spread from the Kitchen house to six other homes on the same block. It then crossed Quincy, burning the next four houses to the east on Cherry as well as

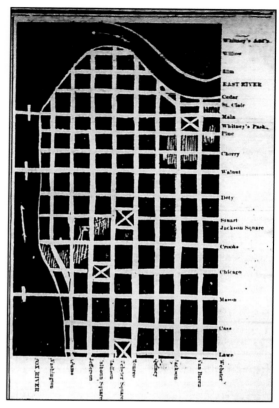

Map from the September 24, 1880, *Daily State Gazette*. The scratched white lines indicate burned areas.

The Great Fire of 1880

Close-up of the northern section of the conflagration area from the 1882 lawsuit map (page 114). Burning debris carried by the wind ignited the roof of the Charles Kitchen house, located immediately beneath the "E" in "Cherry" between Quincy and Jackson. The second, more destructive area of the conflagration expanded from there.

all seven homes on the block immediately south of Whitney Square. A determined fight at the Lutheran church on the northeast corner of Cherry and Van Buren stopped the fire's spread to the east.[61] This building survives to this day as The Bridal Church. However, the fire continued to advance to the northeast, where it "cleaned out" all six buildings on the northwest block of North Webster Avenue and Main Street.[62] From here, the fire died out at Webster and the East River, when "there was nothing more for it to feed upon."[63]

Ironically, the last structure to burn was the fire department dock at the East River on the end of Webster.[64] This dock was 1,500 feet from the Kitchen house and 4,650 feet (nearly nine-tenths of a mile) from the Astor Planing Mill.[65] The second, northern section of destruction "was even more widespread and severe" than the first, southern section.[66]

130   Chapter 6

West De Pere engine house, rigs, and firefighters. This steamer responded to the northern section of the conflagration.

The fire that started at the Kitchen home on Cherry Street occurred at the same time as the successful stand at Jackson Square. Thus, all Green Bay firefighters and equipment were already fully committed. As in the southern section, neighborhood residents in the northern section initially fought the fire. Calls for assistance brought the Fort Howard and West De Pere steamer fire engines to the northern section. These fire companies had already responded to Green Bay.[67] Notably, firefighters saved the Pine Street School House (later the site of Whitney School) on the southwest corner of Pine and Webster.[68] Eventually, GBFD Steamer No. 1 relocated to the northern section, although by then the disaster was essentially over. The

Aerial view down South Adams from St. Willebrord Church in 1891. The wide view (above) and close-up (right, of the upper left portion of the wide view) shows a gap in the trees lining both sides of Adams between Crooks and Chicago. This was the path of the conflagration eleven years before.

conflagration was under control by 7:30 pm, though more accurately, there was nothing left to burn beyond the path of the fire.[69]

The city was devastated by the conflagration. Sixty buildings were destroyed—mostly homes—but also one church, a parsonage, a few stores, as well as about forty other outbuildings, for a total loss of approximately 100 structures.[70] Overall damages were estimated as high as $135,000.[71] In addition, a newspaper story grieved the loss of hundreds of beloved shade trees.[72] Only one woman, Mrs. Schumaker, died of fright as the rain of burning debris threatened her home on Main Street. One firefighter was injured. Henry Reber of the Washington No. 1, suffered a severely gashed foot.[73] GBFD had an overturned and slightly damaged ladder truck, a malfunctioned reel on Hose Cart No. 1, and damaged springs on Steamer No. 2.[74]

Residents had removed a good deal of property, goods, and furnishings from many of the threatened homes as the fire advanced. It was reported that a "small army of people [had] invaded every threatened house, to aid in the work of saving valuables."[75] Much of this was transported to safety. Some valuables piled in the streets burned before they could be moved, while others were lost

The aerial images from St. Willebrord Church previously have been dated to 1889. However, the fall of 1891 is more accurate. There were two consecutive St. Willebrord churches at the same location. From Sanborn Insurance maps, the steeple of the first was about forty-five feet high, while the second (still existing) is 205 feet. These images were taken from a perspective possible only from the higher, second steeple. The *Green Bay Advocate* reported the second steeple was completed in November 1891. It is completely clad with roofing material; there are no viewing platforms or windows. Thus, these panoramic pictures could only have been taken after the supporting structure reached a great height, but before the spire was covered. That was the fall of 1891.

in the chaos.[76] Newspaper notices attempted to return lost and found items to owners for many days afterward.[77]

In spite of the devastation, the fire department quickly received praise. Newspapers stated, "The department put in some excellent work," that there were "well directed efforts at a few vantage points, the sweep of the fire was greatly limited" and the fire department "put in some creditable and efficient work."[78] The papers openly acknowledged that once the fire took hold, stopping the conflagration was beyond possibility.

At a governmental level, Green Bay Common Council meeting minutes made little reference to the fire or the firefighters. Certainly, neither the newspapers nor municipal officials were critical. The *Green Bay Advocate* summarized the praise directed at the firefighters on September 23, 1880:

> The steamers did as well as possible and the Fort Howard steamer rendered efficient aid, but there was too much ground to cover and the water tanks were insufficient for so large a fire.

The substantial praise and lack of criticism can be explained by the fact that conflagrations were not entirely rare in the nineteenth century. Just nine years earlier, the Great Chicago Fire resulted in the destruction of 17,000

Aerial view down South Jefferson from St. Willebrord Church in 1891. The open lots and absence of trees to the left-center are indications of damage from the 1880 fire.

buildings and 300 deaths. On the same night, Peshtigo, Wisconsin, was reduced to ashes with at least 1,000 deaths in the surrounding area. Similarly, Oshkosh, Wisconsin, experienced three conflagrations in less than a year. The last occurred April 28, 1875, and resulted in the burning of several hundred homes and businesses over at least an eighteen-block area.[79] Milwaukee suffered four firestorms from 1849 through 1861. In 1892, the Milwaukee Third Ward Fire resulted in 440 destroyed homes across sixteen blocks. Closer to Green Bay, a major fire on the east side of De Pere in 1882 resulted in the destruction of thirty-six

The Great Fire of 1880

Aerial view toward Jackson Square from St. Willebrord Church in 1891. Firefighters successfully stopped the fire at Madison and Stuart, to the right of the white church with the striped spire. They saved the house at the corner, but eight houses on that same block burned—the open lots to the right-center. The near absence of trees on that side of Stuart (highlighted in the close-up on the opposite page) is the result of the fire storm.

buildings. Even Green Bay earlier had suffered conflagrations in 1841, 1853, and 1863–though these were smaller than the September 20, 1880, fire.

Since urban conflagrations were a reality of the time, it's easy to imagine the residents of Green Bay considered themselves lucky that only sixty homes were lost. In fact, two weeks after the Green Bay conflagration, a local newspaper noted the anniversary of the 1871 Peshtigo Fire and remarked, "In the midst of afflictions, let us be thankful that they are no worse."[80] Much of this "fortune" was attributed to the volunteer firefighters. Consequently, GBFD went unchanged.

There was, however, one aspect of the 1880 fire that generated a force of change to firefighting in Green Bay – water supply. Testimony from several firefighters clearly established the cistern near the fire origin at the planing mill on South Washington was essentially dry due to low river levels. The firefighters struggled with the water supply while the fire was still small and limited to the shaving pile outside the planing mill. During their futile efforts to secure a water source, the firefighters witnessed the flames spread to the planing mill structure and beyond, triggering the firestorm. One firefighter, in particular, was certain the conflagration could have been prevented but for want of sufficient water supply. Within a few days of the fire, a newspaper stated, "The [underground] tanks were insufficient for so large a fire."[81] Below-ground cisterns and the rivers were the only sources of water for firefighting in 1880. Clearly, this system was inadequate and an alternative was needed.

While The Great Fire of 1880 was certainly devastating, the disaster could have been much worse. Just two blocks north of the Astor Planing Mill, the extensive commercial and business district thrived on Washington Street and the adjacent areas. If the wind that day had been just a bit more from the south, the path of destruction would have cut through the economic heart of Green Bay. Consequently, there was a substantial community incentive to take action in preventing another even more economically disastrous fire.

GBFD was not directly altered in the aftermath of the 1880 fire. However, motivation to prevent a reoccurrence prompted creation of a municipal water works, which was a landmark improvement to firefighting in Green Bay.

Not surprisingly, numerous lawsuits were filed against the Goodrich Transportation Corporation, owners of the *Oconto*. Homeowners and insurance companies initially prevailed at least twice, but decisions were overturned on appeal. The February 14, 1889, *Green Bay Advocate* reported yet another trial had been called off and no more were expected, essentially due to costs. It seems Goodrich was never held financially accountable.

The *Oconto* sunk in 1886 after striking a shoal at the western entrance to the St. Lawrence Seaway in New York.[82]

Headline from the September 23, 1880, *Concordia*. The German language newspaper in Green Bay proclaimed "A Great Fire. A Sad Day For Green Bay."

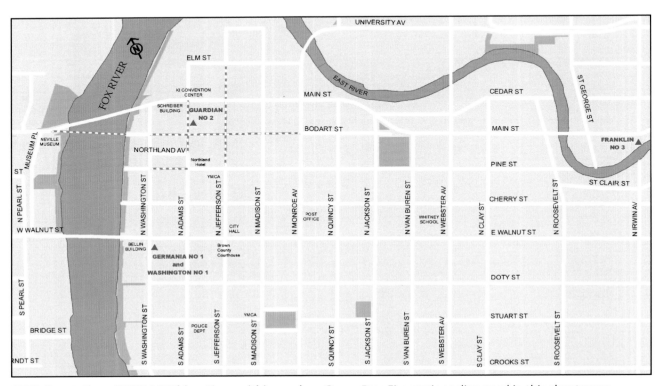

GBFD fire stations (1886-1889) locations within modern Green Bay. Fire stations discussed in this chapter are indicated by the triangles and labeled with the company names. Streets that no longer exist as of 2016 are indicated by dashed lines.

Chapter 7

Green Bay and Fort Howard Water Works Company
1886 - 1889

Municipal-wide water systems became common in larger US cities during the second half of the nineteenth century. Before then, communities retrieved domestic water from individual or communal wells, and water for firefighting came from natural water bodies or underground tanks, known as cisterns.

Municipal water systems provided, safe, potable water, built mostly in response to the realization that waterborne sewage contamination was a frequent cause of disease. An additional advantage was pressurized water distributed through underground mains and accessible at above-ground hydrants—a vast improvement for fighting fire.

"The Water Works doubtless prevented a much more serious fire."

- *Green Bay Advocate*

The issue of a water system had been discussed for years in Green Bay and its cross-river neighbor, Fort Howard.[1] It was not until May 1886 that the prominent citizen A. C. Neville brought a recommendation to the Green Bay Common Council. Speaking on behalf of the very influential Green Bay Business Men's Association, he asked the council to seek bids for a water works.[2] A three-alderman committee was formed immediately. They reported a few months later that five bids had been received and recommended a citywide vote on the most favorable option.[3] On the west side of the Fox River, the Fort Howard Common Council also contemplated building a municipal water system.[4]

In addition to the desire for safe drinking water, fresh memories of the Great Fire of 1880 provided motivation to build a water supply system. That fire began with a small blaze in the wood shaving pile outside a planing mill on South Washington Street. Green Bay Fire Department (GBFD) Steamer No. 1 arrived promptly, but the cistern utilized for water supply in that area, fed by a pipe directly from the Fox River, was essentially dry.

Lack of rain and an especially strong wind pushed water out of the river and into the bay, lowering water level below the pipe's intake opening. Access from a dock was

North Washington with Cherry in the foreground in 1886. The Great Fire of 1880 would have swept through this congested area if the wind had been a little more from the south. Even the all-masonry exteriors required by the ordinance would not have withstood such a massive blaze. Fear of another conflagration, especially through the economically vital downtown, motivated business and civic leaders to develop a municipal Water Works system.

equally ineffective, also due to the low river level.⁵

Because of ineffective firefighting efforts and a strong wind, the flames extended across South Washington, culminating in a devastating firestorm. The business district in downtown Green Bay was spared, but only because of fortuitous wind direction. Fear of another conflagration, particularly through the commercial center, motivated the Business Men's Association to approach the Common Council about a Water Works. The city needed a reliable source of water for firefighting.

Green Bay and Fort Howard Water Works Company   143

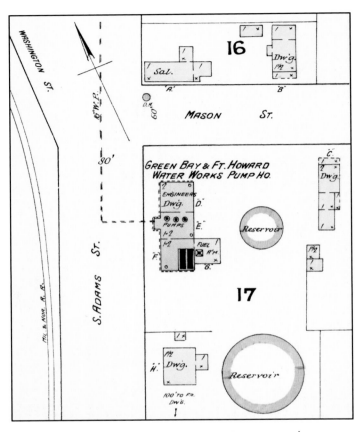

Green Bay and Fort Howard Water Works Company site from the 1887 Sanborn Insurance map. The facility at East Mason and South Adams streets included the first artesian well, three steam-powered pumps, and aboveground reservoirs. Flow from the well could not meet the massive demands of a large fire suppression operation, so the system included reservoirs holding two million gallons of water exclusively for fighting fire. A single sixteen-inch pipe (dashed line) carried water to both cities.

On July 15, 1886, the citizens of Green Bay approved the water works proposal by a vote of 911 for and 42 against.[6] The Fort Howard Common Council passed a resolution to establish a joint water works with Green Bay shortly afterward.[7] With approval from both communities, the Green Bay and Fort Howard Water Works Company was created.[8]

Both Green Bay and Fort Howard passed ordinances defining the Water Works system in great detail.[9] The water would come from an artesian well at a pumping station in Green Bay. The water in an artesian well is under natural pressure and thus forced toward the surface. In Green Bay, this pressure was found to be substantial. Two large pipes would cross the Fox River to supply Fort Howard. Green Bay would have eleven miles of mains and 125-175 hydrants, each with two, 2½-inch outlets. In Fort Howard, there would be six miles of mains and 60-70 hydrants. Service would be provided within the congested commercial centers and adjacent residential neighborhoods. Within these areas, there would be one or two hydrants at every intersection.[10]

Water pressure would be constantly maintained at

40 pounds per square inch (psi) for domestic use and increased to 100 psi "in case of fire." The cities would pay hydrant rental fees amounting to $6,000 per year for Green Bay and $3,000 for Fort Howard. Anticipating municipal growth, the Green Bay ordinance set $37.50 as a yearly rental for each additional hydrant installed in the future. The July 15, 1886, *Green Bay Advocate* reported the proposed pumps would be able to supply two million gallons per day. The Water Works system would be entirely capable of fulfilling demands for consumer and firefighting uses.

Construction on the Water Works began quickly upon approval of the ordinances. Within days of the Green Bay vote, the Water Works company—then a private business—purchased property for the well site and pumping station on South Adams Street at East Mason Street.[11] This is the same location as today's Green Bay Water Utility facility.

The *Green Bay Advocate*, an ardent supporter of the Water Works, chronicled the rapid progress through the summer and fall of 1886. Construction of the facility structure began within a month of purchasing the property.[12] Digging water main ditches and laying water main pipes started in September. It was expected that crews would

THE ORDINANCE PASSED

THE GREEN BAY COUNCIL BY A UNANIMOUS VOTE

Grant a Franchise to the Green Bay and Fort Howard Water Works Co.

Approval of the water works ordinance announced in the July 24, 1886, *State Gazette*. The Green Bay and Fort Howard newspapers universally approved the water supply system and chronicled construction progress.

> *"Pumping engines will be more ornamental than useful."*
>
> - *Green Bay Advocate*

place up to one-half mile of underground main pipe per day.[13] Also in September, the artesian well at the South Adams facility struck water at 400 feet.[14]

By early October, two miles of mains had been placed in Fort Howard and four miles in Green Bay.[15] That same month, crews installed the first hydrant near the D. W. Britton Cooperage on North Monroe Avenue, between Elm Street and modern-day Main Street.[16] The artesian well was completed by late October with a seven-inch diameter shaft reaching 960 feet deep.[17] The Fort Howard mains also were nearly finished.[18]

In early December, all of Green Bay's mains had been connected, the last hydrants placed, and boilers set at the pump station facility.[19] All that remained was to install the pumps, finish a reservoir at the South Adams facility, and place the two supply pipes across the Fox River to Fort Howard.

December 20, 1886, was a milestone day for the Water Works: the large-capacity pumps were finally installed. With these last essential pieces of machinery in place, the connection from the well was opened and—for the very first time—water flowed into the Green Bay side of the system. Up to twenty hydrants were partially opened to purge air.

The system performed as expected, with only a few minor leaks.[20] In fact, the artesian well provided the minimal 40 psi domestic pressure entirely on its own. The pumps were not even needed, prompting the *Green Bay Advocate* on January 6, 1887, to suggest the Water Works "pumping engines will be more ornamental than useful." The entire Green Bay portion was charged with water by early January 1887, just after the date required by the ordinance.[21]

However, the supply pipes crossing the Fox River to Fort Howard still were not complete and complications developed. Supply pipes were necessary because the artesian well and pumping station were located in Green Bay. Consequently, a primary 12-inch pipe was laid on the bed of the Fox River at Mason Street and a secondary 10-inch pipe at Main Street. Engineers developed an ingenious method to contend with the uneven contour of the river bed. Using probes lowered through small holes in the ice covering the river, crews were able to precisely determine the river bot-

Green Bay and Fort Howard Water Works Company facility from the 1893 lithograph. Only one above-ground reservoir tank is shown, though there were two.

View south toward the intersection of South Washington and South Adams in the early 1900s. The smoke stack and adjacent building in the top-center are the Water Works facility. Travelers had to pass over the railroad tracks to reach the Mason Street Bridge to the right.

tom profile. These measurements were then used to shape sections of supply pipe to match the river bed contour and, hence, fit perfectly. Flexible joints between the twelve-foot sections of supply pipe allowed "for the inequalities of the bottom." [22]

This process began in the early winter of 1887. The form-fitted supply pipes were lowered by rope to the riverbed through a narrow gap cut in the 2- to 3-foot thick ice spanning the width of the river. Once laid from shore to shore, the supply pipes were connected to the network of mains and charged with water pressure.

Unfortunately, in spite of the best engineering efforts, this project was beset with problems. The newspapers reported at least four instances when the process was unsuccessful. Each time, the pipes were pulled back to the surface for repair.[23] Finally, in February and March 1887, first one, then the other of the supply pipes were laid across the bottom of the Fox River without complications. The system was opened, and pressurized water was established in Fort Howard.[24]

By March 1887, water supply was, for all practical purposes, in-service in both Green Bay and Fort Howard. Both city ordinances required a formal test for acceptance. However, disagreement between the Water Works company and the City of Green Bay on the terms of the test led to a delay.[25] In reality, it was an actual fire that provided the circumstance which clearly demonstrated the value of the Water Works system.

In May 1887, a barn on Cherry Street near modern Baird Street caught fire. Strong winds extended the flames to another barn and two dwellings. GBFD promptly responded, and for two hours utilized six attack hoses directly attached to three new hydrants. The steamer fire engines were not even used.* Newspaper accounts

*This differs from modern practice. Today, Green Bay firefighters connect a large, five-inch diameter hose from the hydrant to the engine. Fire attack hoses, pre-connected to the vehicle, are deployed to the fire. A pump in the engine provides 150 psi or more water pressure..

SAVED BY WATER WORKS

The First Use for Fire Service in the City of Green Bay.

A Fire on Cherry Street in which the Barns and Residences of E. g. . Woodruff and W. K. Ansorge were Destroyed. The Fire Confined.

Fire report headline from the May 25, 1887, *State Gazette*. Even before the Water Works underwent formal acceptance tests, GBFD successfully fought a barn fire on Cherry Street using hydrant-supplied water for the first time. Firefighters prevented a conflagration despite conditions similar to those of the Great Fire of 1880.

proclaimed that a conflagration was prevented in spite of conditions being very similar to those which led to the Great Fire of 1880.[26] The May 19, 1887, *Green Bay Advocate* stated, "Numerous fires were started by the flying coals, but the Water Works held the key to [controlling] the situation."

Additional statements noted, "This could not have possibly been done but for the services of the plant," and "the firemen and citizens who witnessed this first use of the Water Works plant here for fire protection are enthusiastic in its commendation."[27]

Regardless of the practical demonstration on Cherry Street, official acceptance tests were still required. The entire infrastructure was finally finished with completion of a second, larger, ground-level tank (giving a total reservoir capacity of one million gallons).[28] Fort Howard performed a rigorous test on June 22, 1887, with completely acceptable results.[29] Green Bay also conducted a successful acceptance test about a month later.[30] These tests were described as "much greater than would ever be required at a fire."[31]

Supplemental tests in Green Bay further demonstrated the substantial capabilities of the new water supply system.[32] In one instance, at a location 2.75 miles from the Water Works facility, a pair of 150-foot hose lines were deployed from two hydrants. Water streams reached 145 and 160 feet.[33]

Based on these tests, the *Green Bay Advocate* concluded on July 28, 1887, that, if needed, ten streams could be supported for seven hours and would reach the highest building in Green Bay. Through practical fire suppression and formal tests, it was clear that Green Bay and Fort Howard now had an exceptionally capable water supply for firefighting.

But as with any new, complicated system, officials soon identified problems. The most significant was the means to signal for increased water pressure to support firefighting operations. The ordinances required routine domestic pressure of 40 psi, with increased pressure to 100 psi in case of fire. Additional pumps at the South Adams facility were needed to provide the 100 psi fire pressure.[34] However, it was not possible to isolate increased pressure to one area; it was provided throughout the entire network of mains on both sides of the river.

Most significantly, the ordinance did not specify the means to actually alert the Water Works staff of the

WATER WORKS ACCEPTED.

Test Satisfactory to the Fort Howard Council.

Water Thrown for Three Hours.—The Inspection of the Work at the Pumping Station and at the Nozzles.—Notes of the Test.

Headlines announcing separate acceptance tests of the Water Works system. From the June 19, 1887, *State Gazette* (Fort Howard test) and July 28, 1887, *Green Bay Advocate* (Green Bay test).

The Water Works Test in GREEN BAY.

Triumphant Result of one of the Severest Tests ever concocted—The Plant Proven to be about Perfection.

Green Bay and Fort Howard Water Works Company

need for increased pressure. Recognizing this deficiency a month after the acceptance test, the Fire Department Committee of the Green Bay Common Council proposed a telephone connection between the fire station and Water Works.[35] This does not appear to have been acted upon, most likely due to cost. Two months later, the Green Bay Common Council again proposed a telephone connection to the Water Works. This time, disagreement with the Fort Howard mayor over cost derailed the proposal.[36] A telephone connection to the Water Works eventually was in place by the end of 1887.[37] The reliability of this early system is unknown.

In addition to the telephone, firefighters and Water Works staff relied upon the sound of the fire alarm bells at the engine houses. However, the bells at the two Green Bay stations (No. 1 on South Washington at Walnut, and No. 2 on North Adams at Main) could be difficult to hear at the South Adams Water Works facility. Hearing the fire alarm bells across the river at Fort Howard's engine house was even more difficult.[38] These rudimentary alarms eventually were superseded in Green Bay by a citywide fire alarm system in 1892, which completely solved the problem of alerting the Water Works staff to the need for increased water pressure.[39] That same year,

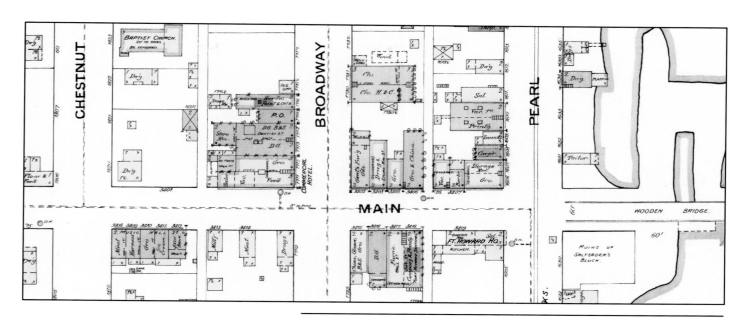

Downtown Fort Howard from the 1887 Sanborn Insurance map (above). Both cities joined to form a common Water Works system. This section of Main Street (today West Walnut) in Fort Howard is served by four hydrants over two blocks. (Water mains are indicated by the double-dashed lines and hydrants by the blue circles.)

Downtown Green Bay from the 1887 Sanborn Insurance map (right). A massive fire destroyed this block in November 1863. Nearly a quarter-century later, seven hydrants encompass this same block. (Water mains are indicated by the double-dashed lines and hydrants by the blue circles.)

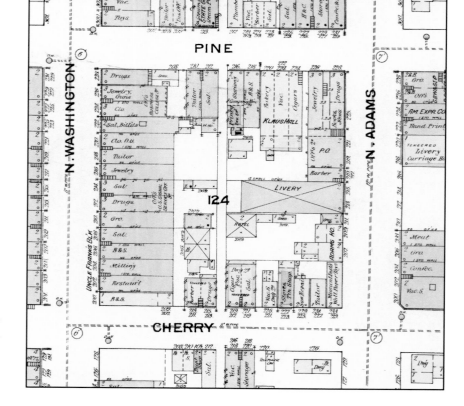

Green Bay and Fort Howard Water Works Company

A. C. Neville home on the southeast corner of the intersection of South Monroe and Porlier in about 1900. The home was built in 1890 (still standing in 2016), while the hydrant in the right foreground was installed between 1890 and 1900. It features two 2½-inch outlets. Notice the wooden plank sidewalks.

Fort Howard also installed a reliable alarm mechanism from the engine house to the Water Works.[40] Overall, the water supply system proved overwhelmingly successful. Shortly after acceptance in Green Bay, a delegation from Waukesha (in Southeast Wisconsin) inspected the Water Works in contemplation of their own and was profoundly impressed.[41]

The best indicators of success, though, came at actual fires. Damage was limited to only part of a house in

Hydrant at the southeast intersection of Walnut and South Washington in the mid-1910s. Zaben Brothers Confectioners is the business on the corner.

an October 1887 fire. The newspaper reported the use of three water streams from nearby hydrants, and that without these, the entire building would have been lost.[42] At another fire near the confluence of the Fox and East rivers at the foot of North Washington, two large businesses were lost, but GBFD saved three others. This was accomplished using hoses attached to the nearest hydrant, which was three blocks away.[43] The newspaper clearly credited water supply as a critical factor in limiting the destruction.

> *"The building stands as a monument to the Water Works."*
>
> - *Green Bay Advocate*
> after firefighters saved a hotel

Similarly, the September 1888 annual review of the volunteer fire department was, in reality, a showcase for the one-year-old Water Works. Eight hose streams, directly attached to hydrants, flowed for nearly two hours while maintaining a system pressure of 90-100 psi. The above-ground reservoir tanks were not substantially depleted in spite of flowing about 150,000 gallons of water, prompting the somewhat-exaggerated statement that the "supply is inexhaustible." Even though this was a review of the fire department, the Water Works eclipsed the firefighters in earning praise with the observation that the "pressure is altogether too heavy for any fire."[44] Many statements regrettably ignored the reality that firefighters were the ones suppressing the flames.

The zealousness with which the Water Works was welcomed resulted in the unfortunate, diminished appreciation for the Green Bay and Fort Howard firefighters. For example, a fire in a hotel on Washington Street was limited to two rooms, with only smoke and water damage to the rest of the hotel. The newspaper stated, "The building stands as a monument to the Water Works."[45] GBFD successfully stopped another fire at Jefferson and Doty streets because "three streams from the Water Works were too much for it."[46] Similarly, a store burned on Main

Street between Clay and modern Irwin. Other structures were in danger, but saved. This prompted the *Green Bay Advocate* to state, "The Water Works doubtless prevented a much more serious fire."[47] The Water Works received all of the credit, with none for the firefighters.

Even more striking, following the fire on Cherry Street in which the water supply system was first utilized, a newspaper account stated, "The fire steamers will rest in the houses hereafter."[48] Apparently, it became common practice to directly attach the attack hoses to the hydrants. The 100 psi fire pressure in the water system was sufficient, thus bypassing – and making obsolete – the steamer fire engines.

A most revealing report was made following a September 1888 fire on South Washington between Walnut and Doty. The upper story of a brick building was destroyed by fire, though the first floor sustained only water damage. The *Green Bay Advocate* concluded that "without the Water Works, it would all probably have burned down, even with the steamer right across the street."[49] In fact, both Fort Howard and Green Bay discussed taking the steamers out of service or even selling them.[50] However, no action was taken.

> *"Without the Water Works, it would all probably have burned down, even with the steamer right across the street."*
>
> - *Green Bay Advocate*

Green Bay and Fort Howard Water Works Company

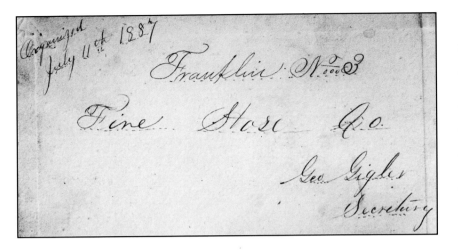

Cover page from the record book of the newly reformed Franklin No. 3 Fire Hose Company. This volunteer fire company formed shortly after the Water Works system was established and served until 1892. Franklin No. 3 was a "hose company" because it had hose carts, but not a fire engine. Firefighters attached the fire attack hose directly to hydrants. Upon a fire alarm, the Water Works charged the entire system to 100 psi, more than enough pressure to provide adequate firefighting water streams. Fire engines were considered obsolete by some citizens at the time.

The Water Works system eliminated some municipal expenses, though hydrant rentals were a substantial cost. The savings was a point used to support passage of the Water Works ordinance.[51] For example, it was no longer necessary to build or maintain cisterns and fire department docks. Also eliminated were costs associated with keeping holes open through the ice on the rivers. The elimination of horse bounties paid to haul the steamers were another major savings. Since the steamers were viewed as obsolete, there was no need to pay to have them brought to fires.

It was further believed the full-time, paid engineers of the steamers had now become unnecessary. A full-time paid engineer was hired each time a steamer was purchased, which was twice in Green Bay and once in Fort Howard. The Green Bay Common Council proposed dismissing one of two engineers three months after the formal Water Works acceptance test.[52] In fact, with the new year of 1888, only one engineer was retained.[53] The Fort Howard Common Council also eliminated its only engineer of the steamer.[54] While the water supply system improved firefighting capabilities, this force of change adversely affected staffing and equipment capabilities of the fire departments in both Green Bay and Fort Howard.

In another aspect, the Water Works actually led to expansion of both the Fort Howard and Green Bay fire departments. The system boosted water pressure to 100 psi in the event of an alarm. Fire hoses directly attached to the hydrants provided very effective streams with this level of pressure, thereby eliminating the need for steamer fire engines. This led to the formation of hose companies in both cities.

The Franklin No. 3 Hose Company was re-established in late 1887 with two hose carts.[55] The first Franklin house, built in 1860, was dilapidated and a new house was built at the same location on Main Street at Irwin.[56] This Franklin house survives today at the Heritage Hill State Historical Park in Allouez. Also at the end of 1887, Fort Howard located a hose cart in a simple shed in the south side of the city. This eventually became

Franklin No. 3 Hose Company house from the 1893 lithograph (above). The city built this house (right-center with a flag atop the tower) at the intersection of Main and modern Irwin, the same site as the previous Franklin No. 3 house (1860-1875). Rahr Brewery is across the street.

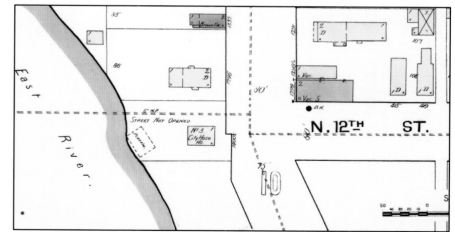

Franklin No. 3 house at Main Street and North 12th Street (modern Irwin) from the 1894 Sanborn Insurance map. Before the water system was built, firefighters placed fire engines on the platform behind the station to obtain water from the East River.

Green Bay and Fort Howard Water Works Company

Franklin No. 3 fire company in about 1890. The firefighters are in their formal uniforms, likely for a parade or the annual fire department review. The East River is in the background. The Franklin No. 3 house survives to this day at Heritage Hill State Historical Park in Allouez.

Resolute Fire Company (discussed in Chapter 10).[57]

Because fire engines were deemed no longer needed, both of these companies had only hose carts. They would attach the hose directly to the hydrants and attack the fire. The water supply system allowed for expansion of the fire departments outward from the municipal centers, thereby improving response time and service to the outskirts of the growing communities.

Although initially there was some overconfidence in the water supply system, the true value of GBFD eventually was demonstrated in November 1892. Repair of a minor water main leak near the South Adams facility required complete water system shutdown for several hours for the first time after several years without an interruption in water service.

During this outage, a barn fire occurred on Stuart Street between Irwin and Baird. GBFD responded, but with no pressure at the hydrants, the steamers could only draw water from private wells and two small cisterns. All were eventually emptied. The fire also destroyed a second barn, which could have been saved if the water system had been operational. Nonetheless, nearby dwellings were protected only because of the efforts of GBFD using minimal water supply.[58] This event demonstrated that the Water Works was a tool of the fire department, not a substitute.

Hose carts in the Franklin No. 3 house at Heritage Hill State Historical Park. GBFD donated these carts to Heritage Hill in the mid-1990s after sitting in fire department storage. The history of these hose carts is unknown, though these would be similar in design to the two hose carts known to have been used by Franklin No. 3.

The benefits of the water system brought about its expansion in the years following initial installation. The common councils of both cities (and the unified city after 1895) frequently received petitions and motions to extend fire mains. Initially, there were 241 hydrants, providing one or two hydrants at every intersection in those areas with water service.[59] Additional hydrants were installed on a regular basis thereafter, with at least one new hydrant placed nearly every year through 1905 and almost one hundred new hydrants installed from 1906-1910.[60] Indicative of the demand, the Common Council established a Water Supply Committee in April 1895 to contend with the frequent requests for water service.[61]

Assistant Chief Henry Reber (left) and Chief Engineer Charles Pfotenhauer in 1886. Firefighters selected these two for the GBFD command positions for the year the Water Works facility began construction. They oversaw tremendous change to firefighting in Green Bay. Their helmets and belt buckles are adorned with "GBFD."

The Water Works dug a second well in 1891 to meet the increased need for water, eventually reaching a total of seven wells by 1907 and eleven by 1916.[62] In 1906, a supply pipe larger than either of the original two was laid across the Fox River from Stuart Street on the east side to School Place on the west side.[63] Having learned from the prob-

New Water Works pump house in 1910. This is the same site as the original 1887 Water Works facility with East Mason to the left, with streetcar tracks, and South Adams (dirt street) to the right.

lems of 1887, crews used a dredge to dig an even-bottom trench on the riverbed down to the sand.[64] Additionally, a new pumping station was opened at the same South Adams site in April 1907, providing substantially increased capacity.[65] The dual system continued with 40 psi domestic and 100 psi fire pressure on demand. Overall, the Water

The new Green Bay Water Works facility from the 1907 Sanborn Insurance map (right). The massive above-ground reservoir tanks holding water for fighting fire were now in a covered building.

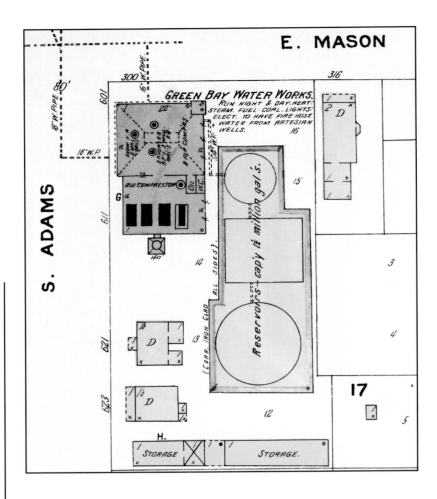

NEW PUMP IS BOON TO CITY

Water Company's Powerful Machine So Viewed by Mayor Minahan.

PRIVATE LOSSES WILL PROVE PUBLIC GAIN

Celebratory headlines from the April 24, 1907, *Green Bay Semi-Weekly*. The 1887 equipment had become inadequate so the water company placed new, larger pumps in an entirely new facility. The comment about private losses refers to the fact that the Water Works facility drew water from deep underground wells, diminishing the water available for the many private wells still in Green Bay.

Works was a huge success and the first twenty years saw extensive expansion.

In terms of firefighting, the most poignant indicator of the Water Works' success is there has never been another conflagration in Green Bay, despite thousands of fires since the Great Fire of 1880. Civic and municipal leaders

164   Chapter 7

Upgraded water pumps in new 1907 pump house.

were motivated to build a Water Works by dread of a repetition of that devastating firestorm. In particular, they feared the thriving business district would be devastated. These fears were eventually assuaged. Much of the credit can be attributed to the Water Works improvements and enhanced effectiveness of the Green Bay Fire Department.

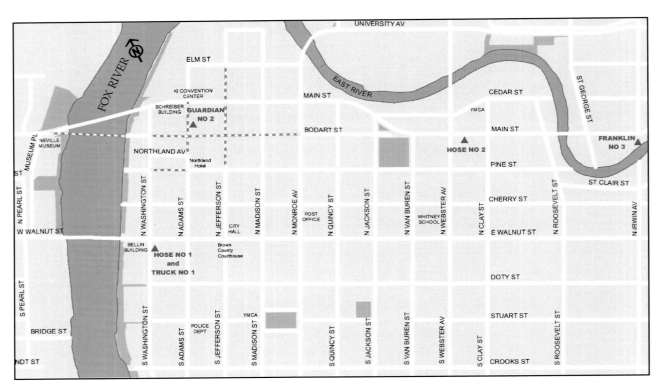

GBFD fire stations (1890-1893) locations within modern Green Bay. Fire stations discussed in this chapter are indicated by the triangles and labeled with the company names. Germania No. 1 and Washington No. 1 volunteer companies used the South Washington station before GBFD became full time. Streets that no longer exist as of 2016 are indicated by dashed lines.

Chapter 8

Birth of the Full-Time, Paid Green Bay Fire Department
1890 - 1893

The birthdate of the career Green Bay Fire Department (GBFD) is May 22, 1891. The place of conception was along the north end of Monroe Avenue at the East River, and the date was Thanksgiving, November 27, 1890. It was at this place and on this date that a fire at the D. W. Britton Cooperage doomed the volunteer fire department, directly leading to the creation of the full-time, paid GBFD.

David Wells Britton (known as "D. W.") was a self-made, typical American entrepreneurial success story. Born in New York in 1832, he came to Green Bay at the age of eighteen and established a cooperage – a business that builds wooden barrels secured with metal hoops, the common storage container of the day. The D. W. Britton Cooperage started in a small shop, then eventually moved

Advertisement for the D. W. Britton Cooperage from the 1884 *City Directory*. A botched response to a fire on November 27, 1890, at this important business doomed the volunteer GBFD and led to replacement with a full-time, paid fire department.

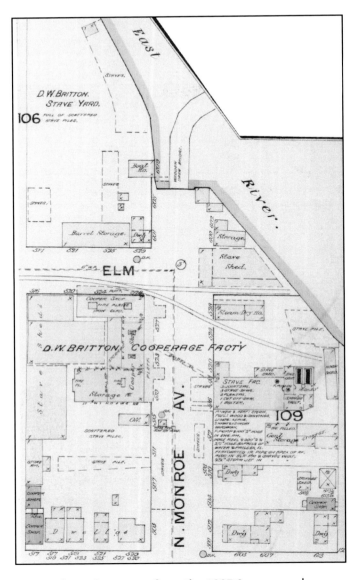

D. W. Britton Cooperage from the 1887 Sanborn Insurance map. The factory was on both sides of North Monroe from modern Main Street (bottom) to the East River. The fire began along the sheds on the left side of the image and destroyed most of the buildings on that part of the block. Cooperage employees initially fought the fire using hoses attached to the hydrant mid-block on Monroe.

to a larger site along both sides of Monroe Avenue, from modern Main Street north to the East River. The company grew into a major Green Bay business, covering nearly three blocks and encompassing fifteen acres, while employing about 130 people.[1] An 1878 newspaper article stated Britton was the second-largest employer in Green Bay.[2]

Besides business, Britton was active in the Green Bay community. He served on the Board of Health and the School Board, was a two-term alderman (also on the Fire Department Committee), as well as a member of the influential Green Bay Business Men's Association.[3] Most significantly, he was a volunteer firefighter with Guardian No. 2 fire company.[4] He joined in 1863 and later served as secretary. Britton was listed in the 1884 *City Directory* as the GBFD Chief Engineer (modern term is fire chief). He was in charge of the fire department for a short time, after which he ended his fire department service.[5]

Unmistakably, Britton was a prominent, well-connected person in the community. Of even greater importance, as a former firefighter, he understood how the fire department functioned – or, more precisely, he knew how the fire department was *supposed* to function.

The Thanksgiving Day fire at the D. W. Britton Cooperage started in a pile of staves, the wooden pieces that comprise the sides of the barrels. The fire soon reached a group of sheds containing more staves and wooden headers, all within the block where Camera Corner Connecting Point is located today. The dry, wooden stock "burned like powder" and spread rapidly. The nearby workshops caught fire and were destroyed. Fortunately, the critical millworks on the east side of Monroe along the river were spared.[6]

Damage to the cooperage business was massive. Much of the facility, particularly the one-block area between Monroe and Madison avenues, from modern Main Street to Elm Street, was in ruins.[7] Britton assessed damages at $25,000 ($640,000 adjusted for 2015).[8] Two adjacent private dwellings also were destroyed.[9]

The cause of the fire was undoubtedly arson. The December 3, 1890, *State Gazette* reported the "incendiary nature cannot be questioned," because eyewitnesses reported the initial shed "flashed like lightening," "the odor of kerosene was detected blocks away," and the "density of the smoke for some time leads to the belief that kerosene was used in large quantities."[10]

David Wells (D.W.) Britton. He was a successful business man as well as former GBFD firefighter and chief. From his fire department experience, he knew the response to the November 27, 1890, fire at his factory was seriously error-plagued.

Britton Cooperage image from the 1889-90 City Directory. This view is from across the East River. North Monroe passes between the large buildings. Those to the right were destroyed in the 1890 fire.

Singularly disturbing was the fact that Britton's business repeatedly had been the target of arsonist attacks. The *State Gazette* dramatically summarized how "after a score of attempts, the unknown incendiary succeeds in burning the cooperage," and that over the previous five months there were attempts "by day and by night to destroy the cooperage of D. W. Britton."[11] This claim was substantiated in the 1890 GBFD annual report. Of the thirty-two total alarms for the year, eight were to Britton's cooperage, including the devastating fire on November

27.[12] Exactly one-quarter of all GBFD alarms in 1890 were to the Britton business. Based on newspaper accounts, there were many attempted fires that did not result in a fire department response. Britton's business was the target of a determined arsonist.

To add to the disturbing reality of being a repeat arson target, Britton, who was present early, witnessed a blundered response by the fire department.

The fire was first discovered at 1:05 pm. Cooperage employees, including Britton, fought the fire using their own fire hose directly attached to a hydrant. In spite of promptly sounding the fire alarm, the closest fire company, Guardian No. 2 only three blocks away on North Adams, did not arrive on scene until 2:00 pm – fifty-five minutes after the alarm.[13] The remainder of the fire department arrived even later, as did the steamer from Fort Howard Fire Department, although the arrival times were not reported.[14]

To make matters worse, the Fort Howard steamer apparently malfunctioned, an allegation later supported by a report to the Fort Howard Common Council and subsequent bills for repair.[15] Britton had arrived at the scene within the first five minutes of the blaze, and as a former firefighter and chief, he had the experience to know the volunteers had botched the job.[16]

FIRE'S RAVAGE.

The Cooperage of D. W. Britton Damaged to the Amount of $25,000.

THE SHOPS WILL BE RE-BUILT.

After a Score of Attempts, the Unknown Incendiary Succeeds in Burning the Cooperage Of D. W. Britten.---The Fire Breaks Out In a Store-House, and Rages for Hours.--- A Man Has a Leg Broken.---Other Incidents.

Disastrous fire reported in the December 3, 1890, *State Gazette*. In spite of the damage, the cooperage was back in limited operation within two weeks.

Mismanagement of the water supply was an even more significant error. Starting in 1887, the Green Bay and Fort Howard Water Works Company provided pressurized water to both cities for domestic and firefighting purposes. The water supply system was maintained at 40 pounds per square inch (psi) for "domestic pressure," but increased to 100 psi for "fire pressure." The original ordinance specifically stated that the Water Works Company "in case of fire alarm, shall with due diligence increase steam and furnish sufficient water so long as needed for the extinguishment of any fire."[17] Unfortunately for Britton, the ordinance did not establish the parameters of an appropriate and acceptable "fire alarm" notification to the Water Works.

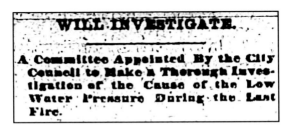

Investigation reported in the December 3, 1890, *State Gazette*. The Common Council quickly appointed a committee to investigate the water supply failure at the Britton fire.

Water pressure was widely reported as entirely inadequate at the Thanksgiving Day cooperage fire. At least twelve water streams were used, but the newspapers reported, "The pressure at the Water Works was so low that the necessary streams could not be produced," and that "the pressure was light and the water had little effect."[18] One account mockingly described how early in the fire, the "ten foot streams were of no use" and shortly thereafter, nearby cooperage sheds caught fire. Britton claimed that well over half of his losses could have been prevented if firefighting efforts had been adequate.[19] His estimate is credible, given he was a former firefighter and chief. Consequently, both the fire department and the Water Works immediately came under close scrutiny.

An investigation into errors at the fire began quickly and with great resolve, an indication of Britton's influence in Green Bay. The very next day, November 28, 1890, the Green Bay Common Council held a special meeting exclusively to "investigate the cause of the low pressure of water at the fire of D. W. Britton's."[20] This special committee reported on December 5 that the machinery was sufficient, "but that the [water] company was unprepared for the sudden and unusual demand made upon it," and most significantly, the "Water Works Company claims not to have received prompt official notice of the fire."[21] A report given to the Green Bay Business Men's Association – the group that originally promoted the municipal Water Works system years earlier – stated the "Water

Guardian No. 2 volunteer fire company about a year before the Britton fire. In the foreground are caps adorned with the company number "2," hose wrenches, nozzles, and the silver trumpet won at the 1860 Brown County Fair Firemen's Competition (page 46). Guardian No. 2 lacked members in 1890 and these six may have been the only company members on the afternoon of the Britton fire. Four of these men went on to be full-time GBFD members: Victor Bader, standing left; Frank Bodart, standing right; Charlie Woodard, sitting left; and John Dewane, sitting right. Sitting in the chair is Isadore Lison who served as Engineer of the Steamer for Guardian No. 2 from 1880 to 1890. At the time of this company portrait, nineteen-year-old Telesphore (Tell) Coel was just made foreman of Guardian No. 2 (standing middle). This image is a family heirloom with Coel's grandsons.

> *"The Water Works were notified of the fire in less than two minutes."*
>
> - D. W. Britton

Works did not put on the necessary fire pressure ... until nearly three o'clock p.m. during which time the fire got a good start and the result was that the large cooper shops were burned."[22] Astonishingly, this established that water pressure was insufficient for almost two full hours.

Britton revealed much of the source of his frustration at a Business Men's Association meeting. He stated, "The Water Works were notified of the fire in less than two minutes. I telephoned them for more pressure."[23] Meanwhile, the position of the Water Works Company was that it did not receive prompt "official" notice, essentially discounting his phone call.[24] Britton was not an authorized municipal official, even though as a business owner he was clearly and substantially involved. Therefore, his request as a civilian for increased water pressure was not followed, as was technically appropriate. Britton thought otherwise. He stated the Common Council report was "a full-fledged 'white wash.' "[25]

Although not specified in the Water Works ordinance, it is reasonable to presume that "official notice" was to come foremost from fire department officials, or alternatively from elected city representatives. A newspaper article written shortly after the Water Works went into operation in 1887 supports this inference. It stated that a telephone was installed at the Water Works facility, but it was to be used only by the fire department in the event of a fire alarm.²⁶ It is reasonable to conclude this policy was in place when the Britton fire occurred three years later.

Blame for the low water pressure had to be placed somewhere. Britton called the Water Works shortly after the fire was discovered. Water Works staff did not respond to notification from a citizen, as was consistent with policy. Water pressure remained low for almost two hours. Every fire department official at the fire scene would have recognized this problem. Exactly when or who finally gave official notice to increase water pressure could not be determined. It is also possible that official notification was made, but Water Works staff failed to comply. Responsibility cannot be determined from existing records, but it remains clear a serious error was made.

The first step in the investigation attempted to determine if the Water Works system was even capable of quickly providing the water supply needed at the Britton fire. Members of the Business Men's Association found an initial report presented at a meeting two weeks after the fire to be frustratingly vague and inconclusive. This prompted an actual demonstration.²⁷ A surprise test of the water supply system was conducted on January 19, 1891 – fifty days after the Britton fire – directly supervised by Mayor James Elmore and GBFD Chief Carl Herrmann.

ALL WERE SURPRISED.

The Mayor and Chief of the Fire Department Spring an Unexpected Alarm to Test the Capacity of the Water Works.

Water Works test reported in the January 21, 1891, *State Gazette*. The unannounced test placed a level of demand on the Water Works system similar to that during the Britton fire. The results entirely satisfied fire department and city officials that the Water Works system did indeed have the capabilities needed at the Britton fire.

The fire alarms were sounded at 7 pm, activating both the Green Bay and Fort Howard fire departments. Twelve water streams flowed simultaneously, same as at the Britton fire. Water Works staff was officially notified without advance warning to increase to 100 psi fire pressure. Water streams flowed continuously for an hour. Chief Herrmann concluded that "the streams furnished were good enough to fight fire with and were satisfactory." Superintendent Harrington of the Water Works Company confirmed that "we were taken entirely by surprise."[28]

This test undoubtedly demonstrated that the mechanical system was capable of fulfilling the demand at the Britton fire. However, blame was never officially assigned. The equipment was absolved, but the person(s) responsible for the failure to either provide or heed official notice was never identified – at least not publically.

Britton was not done with the Water Works Company. He began civil action to recover uninsured damages about the same time as the surprise water system test.[29] His case went to the Wisconsin Supreme Court, which upheld the Brown County Circuit Court ruling against Britton. Very simply, the Water Works Company did not have a contractual agreement with a private business owner. Thus, the company was not liable for damages resulting from low water pressure.[30]

WEDNESDAY, DECEMBER 17.

FIRE DEPARTMENT MATTERS.

A Movement On Foot to Re-organize and Train the Firemen in Their Duties.

Fire department news reported in the December 17, 1890, *State Gazette*. The city implemented steps to improve the volunteer GBFD following the bungled response at the Britton fire.

Some of Britton's resolve to pursue this lawsuit may be explained. First, his business was repeatedly targeted by an arsonist until the devastating fire on Thanksgiving. On a more personal note, his wife of nearly thirty years, Laura Britton, had died on September 1, 1890, less than three months before the fire. She was an invalid for three years and suffered greatly before her death.[31] After dealing with repeated arson attacks on his business and watching his wife die miserably, Britton most likely was in an emotional state and needed an outlet. He was particularly vocal against the Water Works Company at the numerous Business Men's Association meetings following the fire. In contrast, Britton did not make any accusations after a similarly devastating fire at his cooperage in 1878. However, this was before the Water Works existed.[32] Ultimately, Britton had no recourse against the Water Works Company.

Britton, a former firefighter, had witnessed the delayed and error-prone response by the volunteer GBFD. The first fire company did not arrive until fifty-five minutes after the alarm bells sounded, while other companies took even longer.

It is unknown exactly why the response was delayed, but reasonable speculation can be made. Because it was Thanksgiving, it is easy to imagine that the volunteers

Green Bay Mayor James Elmore. He ardently supported creation of the full-time Green Bay Fire Department.

were reluctant to respond, wishing to enjoy the holiday with their families. The owners of the horse teams that were needed to haul the steamers and hose carts were equally disinclined.

Whether because of delayed response by firefighters or horse teams, it is reasonable to conclude that Britton would have been outraged at such a long response time. He fought the fire and watched his business burn during that first hour while knowing the fire department should have already arrived. His anger at the fire department must have been similar to that directed at the Water Works.

In contrast to his vocal criticisms of the Water Works, neither the newspapers nor the Common Council meeting minutes provide any evidence that Britton openly blamed the fire department. As a past GBFD member, Britton may have been reluctant to openly criticize his former comrades. But as a man of political, economic, and social influence in Green Bay, he did not have to be so direct. Whether Britton used his influence or not, it is certain that official municipal attention turned in earnest toward the volunteer GBFD.

The fire department operated under the auspices of the city charter. Thus, the Common Council ultimately controlled the volunteer fire companies. Green Bay

Common Council meeting minutes provide a tremendous understanding of the scrutiny directed at the volunteer GBFD after the Britton fire. On December 5, 1890, about one week after the fire, it was "resolved that the secretaries of the different Fire Companies be requested to report to the council the names of all active members in their respective companies and their attendance at fires during the last year" in order to determine the "actual number of working men."[33] The volunteers were being held accountable.

In an unprecedented event on January 2, 1891, Chief Herrmann directly provided a report to the Common Council.[34] Tantalizingly, the contents of the report and following discussion were not recorded. Even the newspapers, previously complimentary of the firefighters, became critical. The *State Gazette* editorialized that "the recent large fire and the manifest lack of organization of the fire department has renewed the question of re-organization of the department."[35] The paper went on to state, "There has been a decided lack of efficient organization and practical training." One month after the Britton fire, Mayor Elmore and aldermen from the Fire Department Committee went to Appleton and Oshkosh for pointers on reorganization and improvements.[36] Substantial doubts had developed in the volunteer fire department.

"There has been a decided lack of efficient organization and practical training (in the fire department)."

- *State Gazette*

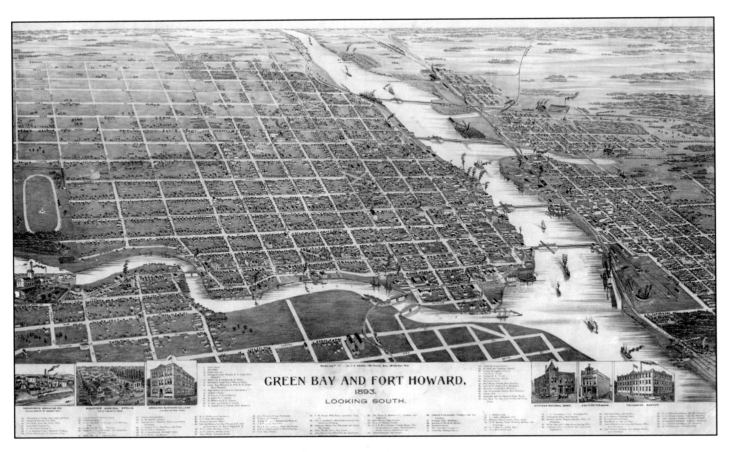

Lithograph of Green Bay and Fort Howard from 1893. A giant-sized version of this image is viewable on a wall inside Titletown Brewing Company, located in the former North Western Railroad Depot just west of the Fox River.

Existing fire company records reveal volunteer firefighter membership was lacking. For example, six months before the Britton fire, Guardian No. 2 consisted of only seven members – four of whom had joined within the previous month.[37] At the same time, Guardian No. 2 meeting minutes discussed a reorganization of the fire company with a determination to "make the Wide Awake Guardian No. 2 one of the most substantial companies of the City."[38] It seems the company had been struggling and fully intended to revitalize.

GBFD consisted of three hose companies and one hook-and-ladder company at the time of the Britton fire. All appeared to have been undermanned. There were only forty-two firefighters between the four companies.[39] Between low membership and the Thanksgiving holiday, it is easy to imagine that not enough firefighters responded in time for a prompt response to the Britton fire.

The city implemented several remedial actions in response to the scrutiny. First, less than two weeks after the Britton fire, the city funded a full-time Engineer of the Steamer for the Enterprise No. 1 steamer. John E. Kittner, a volunteer with Germania No. 1, was hired at $45 per month. Part of the apparatus floor at Engine House No. 1 on South Washington Street was remodeled into living accommodations for Kittner.[40]

There already was one full-time, paid Engineer of the Steamer for Guardian No. 2 at the time of the Thanksgiving Day Britton fire. A full-time engineer for Enterprise No. 1 was assigned in 1868 when the steamer had first arrived, but the position was eliminated in 1888 after the Water Works was completed.

The Common Council directed Chief Herrmann to establish a signal system from the engine houses to the Water Works "without delay."[41] While the mechanism used remains unknown, this was an obvious response to the water supply notification problem that occurred at the Britton fire.

Steps also were taken to improve fire department operations. For example, "plans for re-organization" were established, derived from rules and regulations obtained from leading fire departments in Wisconsin.[42] These rules specified incident command, chain of command, and water supply – details very familiar to modern firefighters.

Next, as a new practice, volunteer firefighters were paid $1.75 for each fire attended – not just a flat yearly rate, as had been done previously. These payments, derived from tax on insurance company premiums, were certainly established to encourage and improve attendance at fires.[43] This was in addition to the two engineers already employed full-time to man the two steamers.[44] In response to these efforts, the February 18, 1891, *State Gazette* reported:

> The department was never in a more harmonious condition than now, and the manner in which Chief Herrmann has taken hold of things and is working for the interest of all companies cannot fail to gain for him the confidence and respect of the firemen and the citizens in general.

Unfortunately for the dedicated volunteers, these efforts were too late in light of the failures at the Britton fire. The February 20, 1891, Fire Department Committee report to the Common Council first used the phrase "paid fire department."[45] This must have angered the volunteers. Within two weeks, members of Germania No. 1, established in 1854, resigned en masse.[46] Shortly thereafter, Washington No. 1 meeting minutes refer to consultation with Germania No. 1 "about division of property held by the two companies."[47] The volunteer system was falling apart.

Chief Herrmann quickly re-established a six-person hose company in place of the resigned Germania No. 1 members.[48] However, the damage had been done. The end of volunteer firefighters in Green Bay loomed.

At the March 6, 1891, Common Council meeting, a report was placed on file "on the plans of a paid fire department."[49] Although the contents of the plans were not recorded, this heralded the beginning of full-time firefighters in Green Bay.

Mayor James Elmore was re-elected in early April 1891 by the "largest majority [79%] ever given a candidate in a local election in this city."[50] He had a mandate. At his April 14, 1891, inaugural address, Elmore lamented "the resignation of Germania Hose Co. No. 1." He stated that reorganization of the fire department "will insure the greatest efficiency," vaguely suggesting a transition to a

Modern alarm box in Buffalo, New York, 2014. Many American cities still utilize this type of system. The bright red color and prominent locations make these boxes easily found in an emergency. The alarm boxes installed in Green Bay in the 1890s had a similar appearance.

A PAID FIRE DEPARTMENT.

The Council Adopts the Report of the Special Committee Recommending a Paid Fire Department.

Celebratory announcement from the May 27, 1891, *State Gazette*. The newspaper reported passage of the ordinance substantially modernizing GBFD. Funding a complete complement of full-time firefighters was the most significant change, making May 22, 1891, a sort of birth-date for the career GBFD.

paid fire department.[51]

Further discussions must have been more substantial and direct, though these were not reported in the Common Council minutes. The aldermen heeded "the request of the Mayor in his inaugural address."[52] The Common Council passed a motion extensively reorganizing the fire department on May 22, 1891. Most significantly, it established full-time, paid positions.

The motion provided a list of changes, but recommended these be accomplished in steps to avoid "too much expense at the start."[53] These changes were as follows:

- Adoption of a municipal fire alarm system
- Discontinuance of Hose Company No. 3 (Franklin No. 3, on Main Street at Irwin Avenue)
- Sell the old Engine House No. 2 on North Adams
- Build a new Engine House No. 2 on Main Street, near Webster Avenue
- Employ three full-time companies at the following salaries:
  - Hose Company No. 1:
    - Driver, $40/month
    - Pipeman, also in charge of electrical apparatus, $45/month
    - Pipeman, also chief, $50/month
    - Four call men, $50/year ("call men" were

184   Chapter 8

part-time firefighters who responded only in the event of an alarm)
- Hose Company No. 2:
  - Driver, $40/month
  - Pipeman, also assistant chief, $40/month
  - Pipeman, $35/month
  - Four call men, $50/year
- Truck Company No. 1:
  - Driver, $40/month
  - Steersman, $35/month
  - Ladderman, $35/month
  - Four call men, $50/year
- Purchase three teams of horses with harnesses and bedding, dedicated to the fire department, which were not to do any work on the streets.

The motion passed by a vote of eight ayes and one against.[54] With that, it was done. The full-time, paid GBFD was born on May 22, 1891.

It would be helpful here to clarify some fire company nomenclature. When this ordinance was passed, crews were formally referred to as hose companies instead of engine or steamer companies. The Water Works Company provided up to 100 psi fire pressure in the event of an alarm. This generated enough pressure that firefighters

THE LAST INSPECTION.

Final Turn Out of the Volunteer Department.

Satisfactory Test of the Water Works Witnessed by a Large Crowd—Fort Howard Makes a Clean Sweep—Reunion of the Veterans.

Bittersweet headline from the September 16, 1891, *State Gazette*. The annual review of the fire department was always a combination of official function and social event. Recognizing the transition to a full-time fire department, Chief Herrmann ensured the event would be memorable. This was the last fire department annual review in Green Bay.

could attach attack hoses directly to the hydrants and then deploy the hose to the burning structure. Consequently, the steamer fire engines were no longer needed and likely routinely left at the stations. Thus, the fire companies who extinguished the fires responded only with hose wagons and became known as hose companies. This naming system lasted into the 1920s, when GBFD began utilizing motorized engines.

These reforms of the fire department were met with approval. The *State Gazette* reported the reorganization "is a move in the right direction" and will "undoubtedly receive the endorsement of a large majority of the taxpayers."[55]

The Common Council made clear in the ordinance that change would occur slowly, specifically due to funding. Initial steps took place over the next few months. In June 1891, the council sold old Engine House No. 2 on North Adams, which was built in 1859.[56] A few months later, a bid was accepted to build new Engine House No. 2 on the south side of Main Street, east of Webster Avenue.[57] This station was designed specifically for a full-time department. It included living quarters on the second story, in contrast to the existing volunteer Engine House No. 2, which lacked accommodations.

No more changes occurred through the summer

Old Croc and veteran GBFD volunteer firefighters in front of Engine House No. 1 on South Washington. These retired firefighters paraded with the Old Croc hand pumper as part of the final review on September 17, 1891. D. W. Britton is one of the veterans, though he cannot be specifically identified.

and fall of 1891. However, an annual city event marked the change from volunteer to full-time GBFD. Each year, the volunteers conducted an "annual inspection" that was partly a display of their abilities and partly a social affair. The annual inspection of 1891 was designated as the "final inspection of the Volunteer fire department," and Chief Herrmann was determined to have "a grand final parade and inspection" as a "fitting finale to a half-century of faithful service."[58]

On September 17, 1891, the sendoff began with a parade from old Engine House No. 2 on North Adams, proceeding through the major streets of both Green Bay and Fort Howard and ending at Engine House No. 1 on South Washington. The parade consisted of a band, followed by the Green Bay Police Department, and then the entire fire departments of both Green Bay and Fort Howard, including the two Green Bay steamers, several hose carts, and the hook-and-ladder truck. Most prominent, poignant, and appropriate was that the firefighters were led by the ancient hand pumper Old Croc, first used in 1841, and accompanied by former volunteers, including D. W. Britton. A photograph (previous page) captured Old Croc surrounded by the veteran volunteers in front of Engine House No. 1.[59]

The fire companies were inspected by the mayor and aldermen. The companies then held a combination "first to water" race and demonstration of the water supply system. At a signal, all the hose companies raced to designated hydrants to be the "first to flow water." A Fort Howard hose company won. At another signal, the Water Works boosted system pressure, and for an hour the spectators were treated to a display of ten tremendous water streams.[60]

Afterward, all the firefighters returned to Engine House No. 2 on North Adams for several speeches. Dr. C. E. Crane, the oldest former Green Bay volunteer, spoke about the early history of the fire department, as did Fort

Howard Chief Engineer A. L. Gray.[61]

Mayor James Elmore addressed the crowd and spoke in admiration of the decades of service by the volunteers. However, he acknowledged there is a time in "every growing city when the volunteer department must give way to a paid department," and that this final review marked the "last step from an embryo to a complete metropolitan, city fire department." Elmore provided a profound endorsement of a paid, full-time department by stating:

> "The operation of fighting fires has been reduced to a science in America, and the men who would do the best service must be constantly employed for such service, and have special instruction and experience, and instead of being scattered about the city at their various places of business, must be with the apparatus and in readiness to respond to alarms in an instant."[62]

From this quote, it is reasonable to suggest Elmore was something of a father to the full-time GBFD. The day concluded with a party at the magnificent Turner Hall, located at East Walnut Street and Monroe Avenue.[63]

As foretold by Elmore when he commented at the final review, "the complete change to the paid department may not occur for several months."[64] Indeed, it wasn't until two months later, November 1891, that the city finally

> —The foreman of each of the fire companies, including the Hook & Ladder company, have received official notice from the Committee on Fire Department, bearing date of Feb. 8, that their services are no longer demanded and that they are formally disbanded, under the charter of the city.

Fire department news from the February 10, 1892, *State Gazette*. The Common Council disbanded the volunteer fire companies and the full-time firefighters went on duty February 5, 1892. In spite of disbandment, the volunteers continued to serve in ever-decreasing capacities through 1892.

funded salaries for the paid firefighters, and then only for the upcoming year.[65] The designated $5,800 covered the expenses for all staffing proposed in the May 1891 resolution. Because of the delay in converting to full-time status, volunteer firefighters remained active well after the final inspection.

While 1891 was a year of expectation and planning, 1892 was a year of implementation of changes to GBFD. First, William Kennedy was appointed chief on January 15, 1892, replacing Carl Herrmann, who had served for slightly more than a year.[66] Next, the Common Council took two nearly simultaneous actions in early February 1892. The three volunteer hose companies and one hook-and-ladder company were officially disbanded, and at the same meeting, the Fire Department Committee announced the names of the first group of four full-time and five part-time firefighters.[67] Included as the full-time engineer for Hose No. 1 was John Kittner, who had been hired as a stop-gap Engineer of the Steamer No. 1 immediately following the Britton fire a little more than a year prior. With these two acts, GBFD transitioned from volunteer to full-time paid status.

More full-time firefighters were added in the following months. Records directly detailing staffing are not available, but the monthly "fire department salary" entries in the Common Council minutes increased through 1892. This stabilized in the fall at a level consistent with about twelve firefighters—even more than what was recommended in the May 22, 1891, resolution.[68] In fact, twelve full-time firefighters were listed in the 1893 Green Bay *City Directory*.[69] This was moderately more than the nine full-time and twelve call men stipulated in the resolution.

Working hours for these initial full-time firefighters is unknown. The earliest records are from 1899, which show GBFD firefighters worked three 24-hour days in a row, then one day off.[70] It is reasonable to conclude that seven years before, the first GBFD full-timers worked the same or very similar shift schedules.

A great tragedy marred the creation of the full-time GBFD. On February 6, 1892, a fire occurred at a Main Street saloon. Hose Cart No. 1 tipped over while responding, injuring its driver, Hans Mark Hansen. He was one of the first full-timers hired, and died of his injuries four days later.[71] It's unknown if he had been a volunteer firefighter previously, but he was injured the first day on the job and almost certainly while responding to the first fire

> **New Fire Department.**
> Following is a list of the names of the members of the new Fire Department; also office and salary of each:
> With Hose Co. No. 1. and H. & L. Co:
> John E. Kittner, engineer, full paid at $45.00 per month.
> E. C. Kittner, call man, $5.00 per month.
> A. Weise, " " " "
> Wm. Hills, " " " "
> Jas. Church, " " " "
> Wm. Johnson, " " " "
> With Hose Co. No. 2:
> Mark Hansen, driver, full paid, at $40.00 per month.
> J. Dewan, pipeman, " " 35.00 per month.
> Frank Bodart " " " 35.00 per month.

List of newly hired GBFD firefighters from the February 11, 1892, *Green Bay Advocate*. Though engineers of the steamers had been employed since 1868, this was the first group to be hired to form the full-time fire department.

Birth of the Full-Time, Paid Green Bay Fire Department 191

Tragedy Strikes the Fire Department

Fire.

Saturday night, at 12:13 the alarm was given of a fire in Sam Davis' saloon on Main street. Fire company No. 1 turned out with promptness and several of the old firemen chuckled over the fact that the paid company "was not in it all."

The hose cart had turned out in proper time but had tipped over and got out of gear generally.

Dr. Munro is taking care of Martin Hansen who was bruised on his back and hips but not dangerously.

The damage done to the building will be covered by insurance but Mr. Davis will suffer a considerable loss as he carries no insurance.

Died.

MARK HANSEN.

Mark Hansen, the hose-cart driver, who, when the vehicle on Saturday night turned upside down, was injured by the shock more than otherwise, died Wednesday. He was 37 years of age, born Jan. 11, 1855. His parents, Mr. and Mrs. Ole Hansen, and one brother, Ole Hansen, survive the deceased. On Friday afternoon, at 2 o'clock, the funeral will take place from the residence of his brother, 1320 Cherry street, Green Bay.

—The pall-bearers at the funeral last Friday of H. Hansen, the unfortunate hose-cart driver, are Ernest Kittner, A. Weise, Jr, H. Nick, Chas. Woodard, Wm. Johnston, J. Dewan.

Newspaper accounts involving the death of Hans Hansen (also identified as Mark or Martin). The February 10, 1892, *State Gazette* reported that Hansen was injured when the hose cart he was driving overturned. Though his injuries were initially reported as "not dangerous," he likely suffered internal injuries that claimed his life four days later (February 17, 1892, *State Gazette*). This same edition listed six pallbearers, five of whom can be confirmed as GBFD firefighters.

---

for the full-timers. The City of Green Bay paid for the funeral and noted the "death was due to accident and himself."[72] Although speculative, one can imagine that in his exuberance and determination to prove himself and the full-time fire department, he drove the horse team too fast. He likely lost control, causing the accident. Hence, within a week, GBFD had learned a lesson repeated relentlessly in the fire service ever since – a momentary poor decision can have tragic consequences.

In spite of an ominous beginning, the full-time GBFD prospered. The volunteers continued to serve, but their involvement diminished. At the same time, appreciation for the full-time department emerged. A March 5,

1892, *State Gazette* story reported that at a house fire, "the paid department had put out the fire" before the volunteers even arrived. The paper emphasized this story was significant only because, in reality, the fire was insignificant. However, it specifically credited the rapid response by the paid firefighters. Similarly, in a May 1892 fire, it was reported that the fire companies left the stations moments after the alarm.[73] More significantly, the paid firefighters successfully handled a massive fire at a large business on the East River in June 1892.[74]

Early on, the reorganized GBFD repeatedly and convincingly demonstrated its benefit to the community. Nonetheless, the Green Bay Common Council continued to make small payments to the volunteer companies, but only through the fall of 1892.[75] By the end of that year, all firefighters in Green Bay were paid full-timers.

New Engine House No. 2 opened in March 1892 on the south side of Main Street, between Webster and

Fire Station No. 2 on Main Street late 1890s or early 1900s. This station opened in April 1892 and was built specifically for a full-time crew, including living quarters on the second floor. The station served until 1965.

Birth of the Full-Time, Paid Green Bay Fire Department

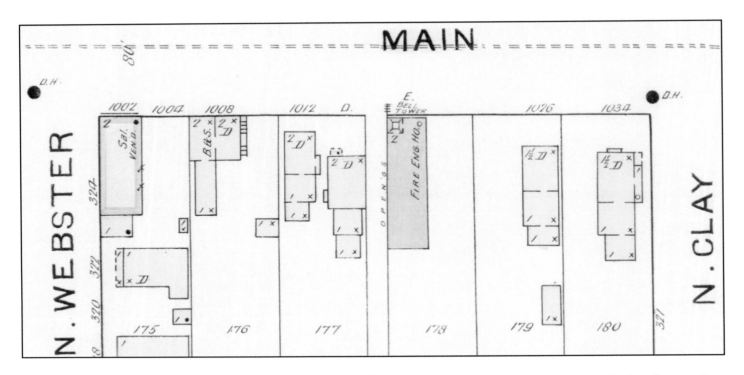

Fire Station No. 2 from the 1894 Sanborn Insurance map. Today, this site (shown in pink in the center) is a parking lot just west of PDQ Car Wash..

Clay.[76] A construction bid was accepted in September 1891, though actual work did not begin until much later. The building finally opened one month after the full-time firefighters began.

The station was designed for a full-time, modern fire department. Quick-opening double doors opened to Main Street from the 1,000-square-foot apparatus floor. Two horse stalls opened automatically when the fire alarm sounded. The hayloft was directly above the stalls and quick-attach harnesses were suspended from the ceilings. Upstairs were living quarters, including sleeping space for five. Also on the second story was a battery room for the new, citywide electrical alarm system under construction at that time. Finally, a two-story tower dominated the front. This fire station remained open until 1965. Retired firefighters recall smelling hay and manure on hot days, even though the horses had been gone since the early 1920s.

About the same time new Engine House No. 2 opened, Engine House No. 1 on South Washington was renovated. The living accommodations were expanded and horse stalls were added on the apparatus floor.[77] GBFD now had two stations suitable for a full-time department.

Horses were purchased for GBFD about the same

Hose Company No. 2 late 1890s or early 1900s. The facepieces on the horses' foreheads are adorned with the number "2" to signify the fire company.

Birth of the Full-Time, Paid Green Bay Fire Department 195

Gamewell alarm box in the Neville Public Museum collection. The stand-mounted alarm boxes were placed most often at street corners. Opening the box door revealed an alarm switch that, when triggered, sent a telegraph signal to the fire stations, which activated gong bells. This box is marked 28, indicating it was at East Mason and Roosevelt streets. It was installed in the early twentieth century.

time as the first full-time firefighters were hired in early 1892. No direct records exist, but a later city expense report clearly lists costs for horse purchases.[78] A May 1892 newspaper account described how "both companies turned out a few seconds after the alarm was given."[79] This could only be accomplished with horses in the fire stations.

A municipal electric fire alarm system was the last fulfilled aspect of the May 22, 1891, Common Council resolution. Representatives of the Gamewell Fire Alarm Telegraph Company demonstrated their system in January 1891, two months after the Britton fire.[80] Discussion followed in subsequent Common Council meetings, but nothing was pursued until the May 22, 1891, resolution specifically mentioned "adoption of some fire alarm system."[81] As with all other aspects of the resolution, the alarm system was delayed until finances were committed.

Beginning in 1892, the city actively sought an electric alarm system and the Gamewell Company again displayed their system for municipal officials and fire department members in January.[82] The system consisted of post-mounted, metal fire alarm boxes at street corners throughout the city. In the event of a fire, a simple pull handle inside the front door of the box would trigger the alarm mechanism. Upon activation, a spring-powered wheel with protruding knobs would rotate and produce a telegraph signal in a specific sequence unique to that box. For example, Box 23 would give a group of two signals in rapid succession, followed by a pause, then a group of three signals. These signals would activate a gong bell at the fire stations in the same sequence. Thus, Box 23 would give two gongs, a pause, then three gongs. This sequence would repeat four times to ensure the firefighters were alerted. In addition, a visual indicator would show the box number, which in the case of Box 23 was at Jefferson and East Mason streets.[83]

To prevent a repeat of the water supply fiasco experienced during the Britton fire, the alarm bell also sounded at the Water Works facility. Thus, Water Works staff knew when the fire department was responding and they would then boost the system to 100 psi fire pressure. This eliminated the need for the official notification that hampered efforts at the Britton fire.

The system was also used for other specific communications. If firefighters on the scene needed more water pressure at the hydrants, they would make three "taps" of the system, then repeat it. This series of three taps would sound at the Water Works facility and alert the staff. Four

Fire Alarm Boxes and Locations.

Box No.		Box No.	
12	Main and Washington.	41	Van Buren and Walnut.
13	Main and Jefferson.	42	Engine House No. 2. Main street.
14	Engine House No. 1, Washington street.	43	Walnut and Eleventh.
15	Doty and Jefferson.	45	Doty and Twelfth.
21	Adams and Crooks.	51	Harvey and Jackson.
23	Jefferson and Mason.	52	St. George and Main.
24	Monroe and Cass.	53	Main and Pleasant.
25	Monroe and Eliza.	54	Harvey and Twelfth.
31	Crooks and Monroe.	61	Murphy Lumber Co. Planing Mill.
32	Crooks and Webster.		
34	Monroe and Cherry.	62	Murphy Lumber Co. Office.
35	Monroe and Cedar.		

Three taps of bell after an alarm, more pressure. 1—1 Fire Alarm Test. 4, Fire out.

To locate a fire, count the blows on the bell as they are struck. Thus: for box 13, 1—3; for box 25, 2—5.

Fire alarm box locations from the 1893 *City Directory*. This is a list of the first twenty-two fire alarms boxes in Green Bay. This street-level system remained in use in Green Bay until the 1950s.

Fire alarm gong bell from a GBFD station in the Neville Public Museum. Signals from the alarm boxes would trigger this bell to sound. The gong would sound in a sequence matching the alarm box number. For example, two gongs then a pause followed by one gong would signify Box 21 at Adams and Crooks.

taps on the system indicated the fire was out and a return to domestic pressure.[84] Gamewell alarm systems were already in place in at least sixteen Wisconsin communities, from as large as Milwaukee to as small as Antigo and Hurley.[85]

The Common Council unanimously accepted the proposal for the Gamewell system shortly after the demonstration.[86] Installation soon began, with the Gamewell alarm system officially tested and found satisfactory on May 27, 1892.[87] A total of twenty-two boxes were placed on street corners in Green Bay. The system cost $4,500 and was paid off over the next two and one-half years.[88]

With the installation of the alarm system, all aspects of the May 22, 1891, fire department reorganization act were fulfilled. There was a new engine house, an exchange of volunteers for full-time, paid personnel, dedicated horses, and an alarm system.[89]

The revamped GBFD faced a major practical test shortly after all facets of the reorganization were complete. Just two days after acceptance of the Gamewell alarm system, there was a fire at a large, abandoned tannery on the south side of Main Street, between St. George and North Irwin (the block that today includes Los Banditos restaurant).[90] Signals from several of the new Gamewell alarm boxes quickly alerted the fire department. The horse-drawn hose cart from new Engine House No. 2 arrived within one and one-half minutes, followed by Hose No. 1 about three minutes later.

The old tannery was a mass of flames, threatening at least ten dwellings and businesses. Some eventually caught fire. The massive Rahr Brewery, across the street at the northeast corner of Main and North Irwin (the site of the present-day Dairy Queen), also was in imminent danger.

Five hose streams were deployed from three hydrants. Initially, but briefly, water pressure was inadequate. Pressure soon increased to 80 psi and ultimately 110 psi, more than enough for firefighting efforts. Under Chief Kennedy's direction, all exposed dwellings were protected. Each developing fire at the adjacent homes was stopped, although the unoccupied tannery burned to the ground. The brewery was undamaged. Some former volunteers assisted, because the firefighting efforts were so considerable. Mutual praise resulted from the interactions

Rahr Brewery in 1895. Located on the opposite corner of Main Street and 12th Street (modern North Irwin Avenue), this huge complex was threatened by the May 29, 1892, tannery fire, but suffered no damage.

> *"Everybody now acknowledges that the paid fire department is just what is required by this city."*
>
> *- Green Bay Advocate*

between the former volunteers and new full-timers. Other American cities endured conflict when full-time firefighters took over from volunteers. Green Bay did not.

Recognition of the service provided by the reorganized GBFD at this fire was resounding. The *Green Bay Advocate* stated, "Everybody now acknowledges that the paid fire department is just what is required by this city."[91] The *State Gazette* noted that Mayor James Elmore and Alderman Frank Murphy, the two men most responsible for the fire department reorganization, were present and:

> The policy which those gentlemen have advocated and carried out was tested on a large scale and under most threatening conditions, and the test vindicated the wisdom of the policy.[92]

Perhaps even more telling was the return of D. W. Britton, whose wrath after the disastrous 1890 Thanksgiving Day fire initiated the drive to reorganize GBFD. On the evening of September 3, 1893, another major fire occurred at the D. W. Britton Cooperage. The enormous stores of wooden staves and headers north of Elm on the west side of Monroe caught fire. A Gamewell alarm box near the cooperage was immediately activated, and GBFD responded promptly. As in 1890, Fort Howard Fire Department responded as well. Eleven hose streams were utilized in an effort similar to the November 1890 Britton fire. Firefighters fought the massive blaze until early the next morning.

The damage was less severe this time. Britton suffered a loss of $8,000, with insurance covering half that amount. As before, an arsonist was suspected.[93] However, unlike the disastrous fire in 1890, the newspapers praised the efforts of GBFD. The Green Bay Common Council did not discuss the fire at all, which was a dramatic difference from three years earlier. Most revealing, however, was the fact there were no complaints from Britton himself.

Firefighting in Green Bay had changed into a capable, modern profession meeting the approval of the community.

GBFD battles a fire at the Green Bay Planing Mill in 1894. This view is facing north on Adams at the intersection of modern Main Street, current site of the KI Convention Center. The fire attack hose is directly attached to a hydrant in the foreground, next to horse-drawn Truck No. 1. The spectators are obtrusively close, a situation familiar to modern GBFD firefighters.

Birth of the Full-Time, Paid Green Bay Fire Department

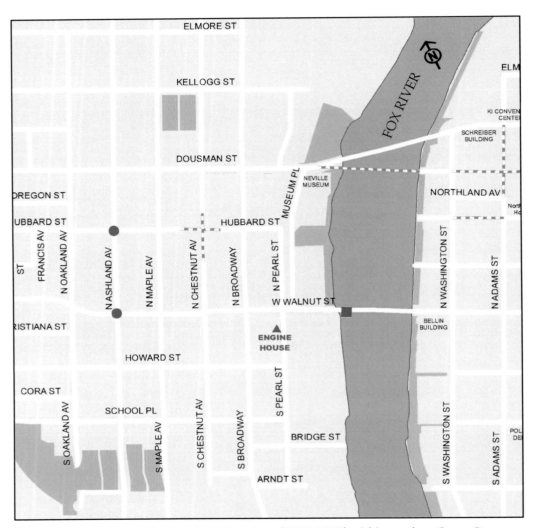

Fort Howard Fire Department-related locations (1860-1873) within modern Green Bay.
The engine house is indicated by the triangle. Water supply sources are indicated by the circles for cisterns and the square on the bridge for a river access hole. Streets that no longer exist as of 2016 are indicated by dashed lines.

Chapter 9

Creation of the Fort Howard Fire Department
1856 - 1873

This historical examination of Green Bay and its fire department must, for completeness, include the municipality of Fort Howard, which today comprises the near west side of the city. Fort Howard began as a small settlement scattered around the military fort on the west bank of the Fox River. It developed into a flourishing community eventually annexed by Green Bay in 1895.

This location was important during early western exploration into North America. The Fox River was a tremendously important waterway connection between the Great Lakes and the Mississippi River, via a short portage to the Wisconsin River mid-state. With Green Bay on the east side and Fort Howard on the west side, the mouth of the Fox River was a strategic point to monitor, protect, and regulate.

Two prominent American politicians served with the US Army at Fort Howard: Zachary Taylor and Jefferson Davis.

Fort Howard US military garrison. The government placed this fort on the west side of the Fox River in 1816 to stabilize what was then the western frontier. The garrison promoted adjacent municipal growth, which eventually became the Borough of Fort Howard in 1856.

The French were the first westerners to govern what is now Northeast Wisconsin. They established Fort St. Francis in about 1687 on the west bank near the mouth of the Fox River. French-Canadian settlers, especially fur traders, colonized the area around the fort. After the French and Indian Wars, the fort was abandoned and control of the area reverted to the British. They rebuilt the fort in 1761 at the same site. It was renamed it Fort Edward Augustus, but abandoned within a few years.

The emerging United States assumed governance of the area after the Revolutionary War. However, control was not asserted until 1816. A US military contingent arrived by ship and established Fort Howard, named after General Benjamin Howard, a commander in the War of 1812. The US soldiers rebuilt the fort on the same spot as the French and English forts, just north of today's Titletown Brewing Company. The river bank was much further inland than at present, having been filled in over the years to its current appearance. The fort and soldiers anchored US presence at this western frontier outpost, stabilized the area and promoted settlement.

Two prominent American politicians served with the US Army at Fort Howard: Zachary Taylor, a commander who later became the twelfth president of the United States; and Jefferson Davis, the only president of the Confederate States of America. Davis was stationed elsewhere in Wisconsin, but visited Fort Howard during his tour of duty.[1]

By the early nineteenth century, the west side of the Fox River had flown three different flags: French, British, and US. The population reflected the same diversity, because civilian settlers followed the military to the frontier. The area flourished through trade in spite of the different languages and cultures. Early fur traders, mostly French or English, shipped products back east from the simple docks along the Fox River. Relations with the later-arriving American soldiers prospered, particularly as the French and British nationals gradually became naturalized. Some Americans from the East and European immigrants moved to the early frontier community. The first steamer carrying settlers from the East arrived in 1821,

By the early nineteenth century, the west side of the Fox River had flown three different flags: French, British, and US.

initiating a great influx of people into the area.[2] Initially, fur was the major commodity. Lumber was the next major industry, fulfilling the demands of a growing nation in the 1800s. Shipping also flourished, transporting products from Green Bay and Fort Howard.

Municipal organization became necessary as the area on the west side of the Fox River developed. The community held a first meeting in April 1842 and a town board formed.[3] More formally, the Borough of Fort Howard was incorporated in October 1856.[4] The one-mile-square borough was along the southern edge of the fort, and eighty-four votes were cast to decide the issue.[5] The newly formed borough was indeed small.

Northeast Wisconsin was no longer considered "the frontier" by this time. In fact, the fort already had been abandoned for four years. Subsequent release of the massive surrounding military reserve to the public triggered great municipal prosperity.[6]

The founders of the borough recognized the need for fire protection. The 1856 "Act to Incorporate" included sections on fire protection and established authority to procure firefighting equipment, such as engines, hooks, ladders, and buckets.[7] The law also regulated building con-

Sections from the legal act passed October 13, 1856, establishing the Borough of Fort Howard. These passages gave borough officials the authority to create, equip, and manage a fire department. Some simple building construction regulations were included as well as the duties, responsibilities, and benefits of fire department service.

struction, in particular chimneys and stove pipes, which were frequent sources of accidental fires. In addition, these rules required building owners to keep ladders and fire buckets available, and to safely deposit ashes. Lastly, the articles established fire warden positions. Although vaguely defined, these officials provided the first municipal fire safety oversight. Fire protection in Fort Howard had begun.

Municipal responsibilities included organization of a fire department.[8] The borough held authorization to establish and regulate fire companies, and disband them as

The borough initially arranged for the hand pumper to be stored in the warehouse of Fisk and Company.

well, but the fire companies were governed internally by their own rules. Fort Howard volunteer firefighters were exempt from jury duty, poll tax, and military draft, except in case of war. These exemptions became permanent after ten years of fire department service.

Unfortunately, very few contemporaneous records describe the early history of the fire department. An 1891 newspaper article included a brief history of the Fort Howard Fire Department, provided by Chief A. L. Gray. He stated, "Howard Fire Company No. 1" was established in 1856, the same year the borough was created.[9] The first formal records discovered are from the May 3, 1858, Fort Howard Common Council meeting, at which a special committee was formed to procure a fire engine.[10]

The Fort Howard Fire Department developed rapidly during the summer of 1858. In June, the special committee agreed to purchase a used hand pumper fire engine from the Department of the Military through Major Ephraim Shaler, a retired officer serving as caretaker of the inactive Fort Howard military post.[11] This is the same Shaler who had sold the Old Croc hand pumper to Green Bay in 1843. In July, voters approved a levy (22 for and 9 against) to raise $500 to purchase the hand pumper and hose from the US military.[12] That same month, the Common Council appointed a fire warden and established a committee to create a relevant ordinance.[13] Howard Fire Company No. 1 took possession of the hand pumper in August, and the Common Council voted to purchase 300 feet of hose.[14]

The borough initially arranged for the hand pumper to be stored in the warehouse of Fisk and Company, most likely on Main Street (modern West Walnut Street), at the northeast corner with Pearl Street.[15] Fisk and Company presented a bill of $5 for five months storage at the end of that year.[16]

To complete the fire department, the Common Council authorized construction of an engine house. The first dedicated fire station in Fort Howard opened on December 21, 1858.[17] Municipal records do not reveal the location of this engine house.

A significant problem with the hand pumper was exposed shortly after purchase from the military. This was the second hand pumper fire engine at the Fort Howard garrison after the first was sold to Green Bay in 1843. This second engine was obtained in January 1850 after the military had re-occupied the fort.[18] The fort closed again in June 1852, this time permanently, leaving Major Shaler as caretaker.[19] It is reasonable to assume Shaler neglected maintenance of the hand pumper during the intervening six years. Immediately after the fire department took possession, it spent $2.94 and $12.00 (substantial sums in 1858) for materials and repair.[20]

The poor condition of the hand pumper apparently was serious enough that Shaler offered to take back the engine shortly thereafter.[21] The Common Council authorized payment to the military of $250 in February 1859, but within a few months the municipal authorities discussed returning the hand pumper. They were clearly dissatisfied.[22] A few months later, in August 1859, the borough gave Fort Howard Fire Department member T. F. Baily $250 (and $25 for travel to Chicago) to purchase a new hand pumper fire engine. The borough paid $15.50 to ship this new apparatus back to Fort Howard.[23] At the same time, it returned the original hand pumper to Shaler at the military post.[24]

Fort Howard Fire Department had a new, reliable hand pumper in the fall of 1859. Original purchase records do not specifically mention the manufacturer. However, a bill for hose in November 1860 suggests Button and Blake as the manufacturer.[25] An 1891 historical review of the Fort Howard Fire Department states the first hand pump-

Button and Blake hand pumper (1855 model) at the Firemen's Association of New York Museum of Firefighting in Hudson. This is the same manufacturer and similar age as the hand pumper purchased for the Fort Howard Fire Department in 1859. There are no known images of the Fort Howard hand pumper.

er was indeed a Button.[26] Having the new hand pumper fire engine firmly established an effective firefighting force in the new Borough of Fort Howard.

Fort Howard Fire Department improved through the next few years, beginning with the purchase of a hose cart soon after the hand pumper.[27] Firefighters deployed the fire attack hose from the hose cart to the burning building. They connected the other end of the hose to the hand pumper, which was always placed at the water source. The Common Council authorized improvements to the engine house, such as banking the front apron and

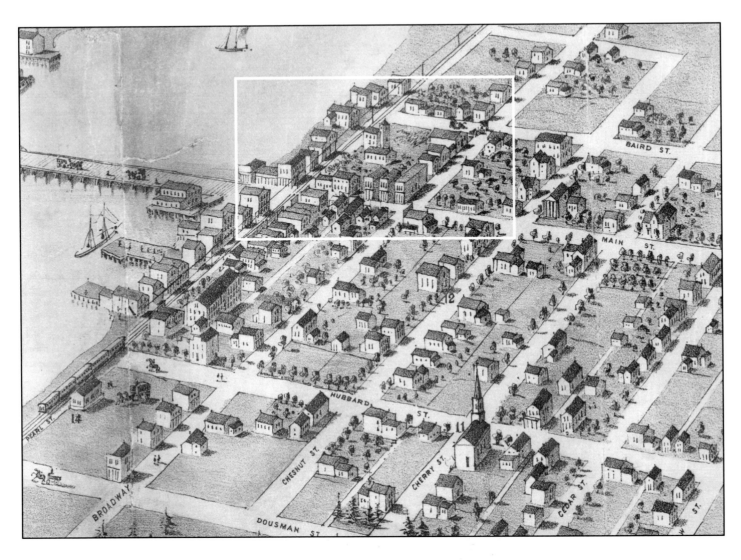

Borough of Fort Howard from the 1867 lithograph. The commercial and mercantile area was on the three blocks centered on Pearl Street and Main Street (modern West Walnut) adjacent to the bridge. The first railroad in the area, built in 1863, passed down Pearl, adjacent to the river. The area within the yellow box is featured in the close-up on the following page. (See full lithograph on page 34.)

installing a metal triangle to sound alarms.[28] Also, the borough purchased a speaking trumpet for the chief engineer (contemporary term for fire chief) to facilitate commands during fire scene operations.[29]

In August 1860, a citizen offered a vacant plot of

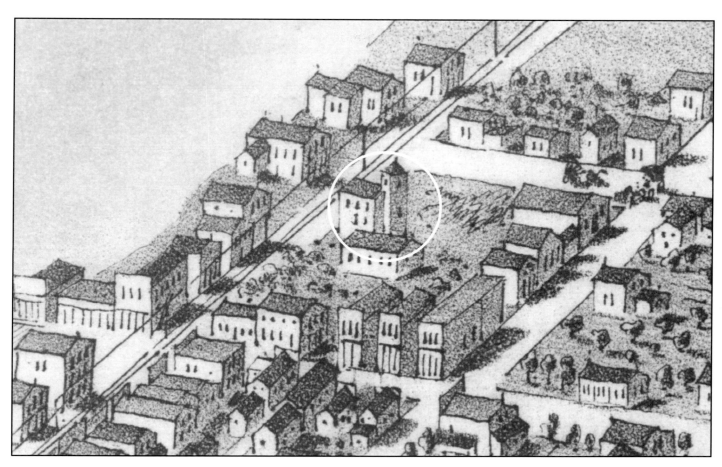

Close-up of the Fort Howard Fire Department Engine House from the 1867 lithograph. The borough built this engine house in 1858, then moved it to the South Pearl location in 1860. Fire destroyed this building in 1872

land for the engine house if the borough would pay the taxes.[30] This lot was located on the west side of South Pearl Street, just south of modern West Walnut. The Common Council accepted this offer and paid to move the existing engine house to this location in the fall of 1860.[31]

Starting that winter, the borough paid men $5 to $12 a month during cold weather for "keeping fire at [the] fire house," to prevent the apparatus and equipment from

Creation of the Fort Howard Fire Department

freezing.[32] The Common Council approved construction of an elevated tower for a new $40 bell to be added to the engine house in December 1863.[33] The Fort Howard fire station had become fully functional.

The hand pumper required a suitable water supply to be effective. Most often, firefighters parked the hand pumper at the river, on the bank, or a dock. They then placed a short supply hose from the hand pumper into the water. Domestic wells served as water sources away from the river. During a December fire, firefighters actually placed the hand pumper on the ice covering the Fox River and accessed water through a hole cut through the ice.[34]

These water sources were not entirely satisfactory. For instance, fire destroyed a barn and stable in spite of arrival of the hand pumper, because there was "no water to be had."[35] Much of the borough was either too far from the river or on-site water supply was inadequate. In reality, parts of Fort Howard were without fire protection.

In October 1867, the Common Council ordered an "approach to the river" built, as well as two access holes on the only Fox River bridge (today's Walnut Street), specifically for fire department use.[36] These must have been successful, because the Common Council authorized

Fire destroyed a barn and stable in spite of arrival of the hand pumper, because there was "no water to be had."

construction of two additional locations with access to the river in early 1869.[37]

At the same meeting, the Common Council approved construction of two below-ground "reservoirs," also known as cisterns, at intersections that are today's Ashland Avenue/Hubbard Street and Ashland Avenue/West Walnut Street.[38] These spring-fed cisterns provided water supply four blocks from the Fox River, what was then the western edge of the borough. The cisterns cost $175 total.[39] Firefighters placed the hand pumper at the underground cistern and lowered a suction hose into the water through a small access opening. Water supply for firefighting was now substantial within the expanding borough.

Along with a fire department to suppress fire, municipal acts also prevented uncontrolled fires. An ordinance passed November 11, 1869, prescribed the "Fire Limits" as the geographical area, essentially the commercial center of Fort Howard, with the following construction requirements.[40]

> All outside and party walls shall be made of stone, brick or other fireproof material, all roofs, cornices and gutters shall be covered on the outside surface with copper, tin, iron or other fireproof material; and all end and party walls shall extend above the sheeting of the roof at least thirty inches.

All outside and party walls shall be made of stone, brick or other fireproof material ...

- Fort Howard ordinance

Basically, wooden exteriors were prohibited. Fire-resistant exteriors inhibited the spread of fire from building to building in the congested commercial area.

Privies (outhouses) with a maximum of eight feet square and eight feet high were exempt, so could have wood exteriors. Because of the dangers from heating and cooking fuels (wood and coal), the ordinance particularly emphasized that ashes must be deposited within brick or other fireproof containers. Fines for violations were set at $20 to $300. Along with firefighting, ordinances suitably addressed fire prevention in the borough.

As occurred in Green Bay in 1868, a major technological development dramatically changed the Fort Howard Fire Department. The borough purchased a Button and Son third class steamer fire engine in 1872, along with two hand carts and 1,000 feet of hose for a total of $5,370.[41] The steamer fire engine, named "Fort Howard No. 1," generated much greater water stream volume and distance compared to the hand pumper.[42]

> **New Steamer.**—The new steam fire engine built for our Fort Howard neighbors by the Button Engine Manufacturing Co., arrived on Friday last. The manufacturer, Mr. BUTTON came with it, it and is superintending setting it up. It is a handsome machine of the third class, and is comparatively light and easy to handle. It is named "Fort Howard, No. 1." It was to be brought out for trial and acceptance on Wednesday (yesterday) afternoon.

Arrival of the Fort Howard steamer announced in the March 21, 1872, *Green Bay Advocate*. Success of the steamer fire engine obtained by Green Bay in 1868 prompted Fort Howard officials to purchase one as well.

The "third class" term described capacity rather than quality, and the Fort Howard steamer could throw two 1 1/8-inch streams at least 200 feet.[43] The steamer used coal as fuel. Consequently, a large contingent of firefighters was no longer necessary to provide power, as was the case with the hand pumper. The steamer required only a few firefighters. The adoption of this technology significantly improved firefighting capabilities of the Fort Howard Fire Department.

The conversion to the steamer provided an insight into relations between the Fort Howard and Green Bay fire departments. Larger American cities frequently sold their outmoded hand pumpers to rural fire departments after obtaining steamers. The Borough of Fort Howard sold its hand pumper to the newly formed Waupacaa Fire Department (spelling accurate for the time of today's Waupaca) in 1871 for $700*.[44]

The timeline of the hand pumper sale by Fort Howard is noteworthy. Records clearly show the Fort Howard Fire Department was without any fire engine at all from the sale of the hand pumper in April 1871 until the new steamer arrived in March 1872.[46] Surprisingly, municipal authorities displayed no urgency toward purchasing a new steamer. Shortly after selling the hand pumper, the

Village of Waupacaa historical records confirm this purchase. Eventually, this hand pumper was again replaced by a steamer in Waupacaa. Beyond this event, the fate of the Button hand pumper, originally with Fort Howard Fire Department, is unknown.[45]

president of the Common Council stated he did "not think purchase of new one . . . advisable at present."[47] In fact, the Common Council did not even solicit bids for steamers until nearly eight months after the hand pumper was sold.[48] A newspaper understated the situation by observing, "The only protection on the front street now is the force pump of F. Blesch [brewery]."[49] It appears neither the municipality nor the community were in a hurry to restore fire protection.

On the surface, lack of a fire engine in Fort Howard seems imprudent, but there is a reasonable explanation. Shortly after Fort Howard Fire Department received its new engine in 1859 they began to respond to fires in Green Bay.[50] This act of assistance, today known as mutual aid, occurred frequently and willingly in both directions across the Fox River.[51] This is borne out by a passage from the 1874 Green Bay and Fort Howard *City Directory*, which states:

> "It is a matter of credit to the different fire companies of Green Bay and Fort Howard that they mutually assist each other and in this manner increase the efficiency of each. Such good feeling and unity of purpose cannot but prove pleasant to their members as also beneficial to the cities."[52]

The Common Council meeting minutes for both Green Bay and Fort Howard do not refer to a formal agreement between the municipalities or fire departments, but it is very likely the Green Bay Fire Department provided fire engine response to Fort Howard when it lacked any fire engine during parts of 1871 and 1872.

Though the steamer provided a vast improvement to firefighting in Fort Howard, it created a significant challenge. Steamer fire engines were much heavier than the hand pumpers. They were mostly metal and held a large volume of water in the boiler. The Button and Son third class steamer used by Fort Howard Fire Department weighed about 5,000 pounds.[53] As with the hand pumper, firefighters could haul the steamer by hand, but the massive weight on dirt streets proved difficult, if not impractical. For example, in January 1875, a

Button and Son steamer fire engine (1871 model) at the Firemen's Association of New York Museum of Firefighting in Hudson. This museum piece is serial number 66, while the Fort Howard Button and Son purchased in 1872 was serial number 75. There are no known images of the Fort Howard steamer. This steamer is in a hand-drawn configuration. Firefighters lowered the upright bar to steer and many others pulled the steamer using rope deployed from the reels between the front wheels.

Creation of the Fort Howard Fire Department

fire destroyed several houses on the southern edge of Fort Howard (near modern Tank School) when the steamer didn't arrive for an hour "having been drawn by hand."[54] Winter weather, combined with dirt streets, made this task much more difficult or even impossible. In several instances, the steamers of both cities became stuck in the muddy streets.[55]

Hand-hauling the hose carts was also difficult. In September 1876, Fort Howard Fire Department responded into Green Bay. After pulling the first host cart over the bridge by hand, the Fort Howard crew returned to get the second hose cart. All the Fort Howard firefighters were needed to pull one hose cart at a time. The streets were so muddy they could not bring both carts simultaneously.[56]

The most obvious solution was to purchase a team of horses dedicated to the fire department. However, this was prohibitively expensive. As in Green Bay, the Fort Howard Common Council implemented a "bounty" system. This program paid citizens or firefighters who were the "first team" to attach their privately owned horses to the steamer and hose carts.[57] Starting in April 1872, the borough paid $8 for the first horse team and $4 for the second horse team to respond. This second team pulled the hose cart. Payment was absolutely dependent on bringing the fire apparatus back to the engine house. Further incentive was provided by the firefighters themselves who offered a $2 supplemental bounty, paid from a fire department fund, for the first horse team from 8 pm to 7 am. Established in April 1875, it is unknown how long this overnight-bonus program remained in effect.[58]

The horse bounty approach had mixed success. The February 22, 1877, *Fort Howard Herald* reported the steamer left the engine house "one minute and a quarter after first alarm given." In contrast, for a fire the next month on Wolf River Road (today's Shawano Avenue), the same team of horses that hauled the steamer had to return to the engine house to get the hose cart.[59] Apparently only one horse team responded. Although the capabilities of the fire department had improved, difficulties persisted in getting the steamer and hose carts to the fire.

The steam fire engine was a complicated mechanism. A coal-fired onboard boiler generated steam that ultimately powered a water pump. Water in the onboard boiler was heated to boiling as quickly as possible. To reduce the "time to steam" in an emergency, the onboard boiler was connected to a water heater when the steamer was parked at the engine house.[60] Quick-disconnect pipes connected the steamer to the water heater on the appara-

> *The steamer left the engine house "one minute and a quarter after first alarm given."*
> - Fort Howard Herald

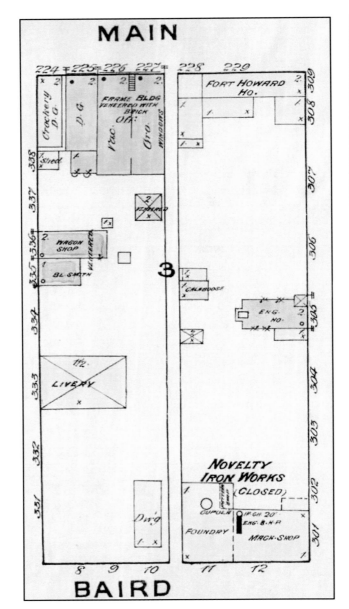

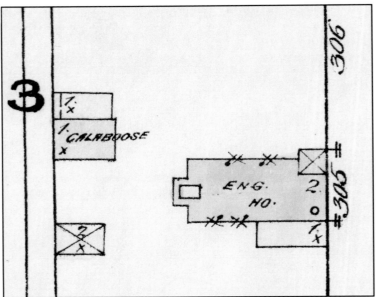

Fort Howard engine house on South Pearl from the 1879 Sanborn Insurance map (right-center of the image on the left and close-up above). This brick exterior building replaced the engine house destroyed by fire in 1872. The bell tower is indicated by the box at the front. Behind the engine house is the calaboose (in blue), another term for a simple jail. Main Street in old Fort Howard is today's West Walnut, and Baird is today's Howard Street. In the left image, South Broadway is on the left and South Pearl on the right.

tus floor, continuously circulating hot water through the steamer boiler. When the alarm sounded, a fire was quickly started in the burn box, further heating the water to a full boil. Having water in the onboard boiler already hot greatly reduced the time needed to produce steam.

Besides hardware considerations, the complicated mechanism of the steamer required substantial operator attention. Accordingly, the Common Council hired Peter Sheridan as Engineer of the Steamer for $900 a year soon after delivery in 1872.[61] Sheridan previously was the Green Bay Fire Department Engineer of the Steamer for Enterprise No. 1 from October 1869 to April 1872.[62] The position in Fort Howard was defined as follows:

> He shall at all times be on duty within one block of the Engine House unless by permission of the Mayor or president of the council and shall be in the Engine House without fail from 12 o'clock noon until 1 o'clock pm. He shall ring or cause to ring the bell on the Engine House at 12 and 6 o'clock daily.[63]

Fort Howard maintained this full-time, paid position for well over a decade, greatly advancing its fire department.[64]

Fort Howard Fire Department endured two controversies during the summer of 1872. At the June 17, 1872, meeting, the Common Council approved "that the old Company be suspended and that the hose and apparatus of the fire company be turned over to the new company." The Common Council formed a special committee to investigate fire department "alleged deficiencies" and "notify them [firefighters] they have the right to be heard in their defense."[65] The Common Council minutes did not describe the events leading to this situation.

The June 22, 1872, Green Bay *State Gazette* reported that at a recent grocery store fire on Broadway in Fort Howard, "the Fort Howard steamer was badly managed on the start but picked up towards the end." Another newspaper stated, "The Green Bay steamer was on hand in quick order and got a stream on the fire very soon," and that "several hand engines did their share."[66] It seems a poor performance by the Fort Howard firefighters, especially in comparison to the cross-river fire companies, created the controversy. This led to dissolution of the existing Fort Howard Fire Company.

A new fire company was formally recognized at the Fort Howard Common Council meeting one week later, given permission to organize, and take charge of the steamer.[67] Other related changes occurred through the summer. In early July, the new fire department members nominated a chief engineer and assistant.[68] These two were elected from the new company and confirmed by the Common Council to replace the two previous chiefs. Peter Sheridan continued as the full-time engineer of the steamer.[69]

At the August 12, 1872, Common Council meeting, a petition was presented regarding the "Live Oak Company."[70] At the same meeting, the aldermen ordered the city marshal to collect from "Howard No. 1 Co. all the property of the Borough in their hands and turn the same over to the Live Oak Company."[71] Live Oak Company was the name of the new Fort Howard fire company. For unknown reasons, meeting minutes used this name only at this session. Common Council meeting minutes never included the name Like Oak Company again.

A last, revealing issue, was a series of bills from just-dismissed Chief Engineer James A. Beattie. He charged the Common Council $380 for "services on Hand Engine in 1866 and for selling of same and purchasing of Steamer." The aldermen unanimously rejected this bill in February 1873.[72] Seeking payment for services seven years earlier strongly indicates animosity. This was the last issue the Common Council addressed regarding the 1872 reorganization of the Fort Howard Fire Department.

In addition to controversy regarding the firefighters and chiefs, the South Pearl Street engine house was destroyed by fire on November 6, 1872.[73] The very next day, the Common Council discussed "what action the Council would take in replacing the Engine House burned on yesterday and to hear the views of the Engineer on the probable cause of the fire."[74] Peter Sheridan testified he had been alone at the engine house all morning and there had not been a fire in the stove the entire summer. He went out for

"The Fort Howard steamer was badly managed on the start but picked up towards the end."
- State Gazette

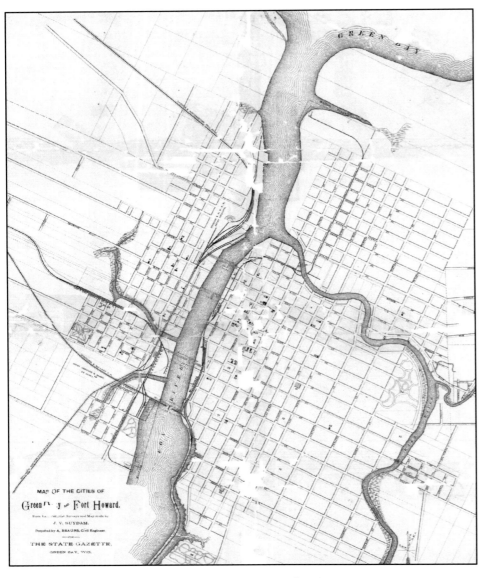

Map of Green Bay and Fort Howard from 1874. J. V. Suydam and A. Brauns prepared this map for the *State Gazette* newspaper of Green Bay. The three bridges over the Fox River are evident, as are railroads in both cities.

five to ten minutes, then heard the fire alarm bell. He thought the fire "seemed to have started under the stairs," where he kept shavings for kindling and dry cotton rags. He testified that he always burned greasy rags after use, discounting this as a cause. Oil and grease on natural fiber cloth will ignite if improperly stored. However, he "thinks it is probable the fire originated from a spark from the Switch Engine," small railroad locomotives used to move railroad cars short distances. The only railroad track through the center of Fort Howard was on Pearl Street, directly in front of the engine house.[75] A newspaper openly mocked this explanation as implausible.[76] Ultimately, a definitive ignition source was never formally reported.

Although the Fort Howard Engine House was destroyed, the steamer was saved with only minor damage to the seat and cushion. Repairs cost $4.75.[77] The steamer was temporarily stored at Green Bay Engine House No. 1 on South Washington Street.[78]

The Fort Howard Common Council quickly pursued rebuilding of the engine house at the same location. Within days of the fire, the Common Council approved $4,000 for reconstruction.[79] The city paid many construction bills through the winter and early spring of 1873.[80] The new engine house opened in early 1873.[81] It was two stories, with a brick exterior and a bell tower. This volunteer Fort Howard Fire Department Engine House became Green Bay Fire Station No. 3 upon annexation in 1895 and served in this capacity until 1937.[82]

Fort Howard Fire Department stabilized after resolution of the two controversial events of late 1872. The fire department was reorganized. Firefighters had a new engine house, a capable steamer operated by a full-time engineer, at least one hose cart, as well as numerous cisterns and river access points for water supply. The fire department had changed, paralleling development of the borough into a city.

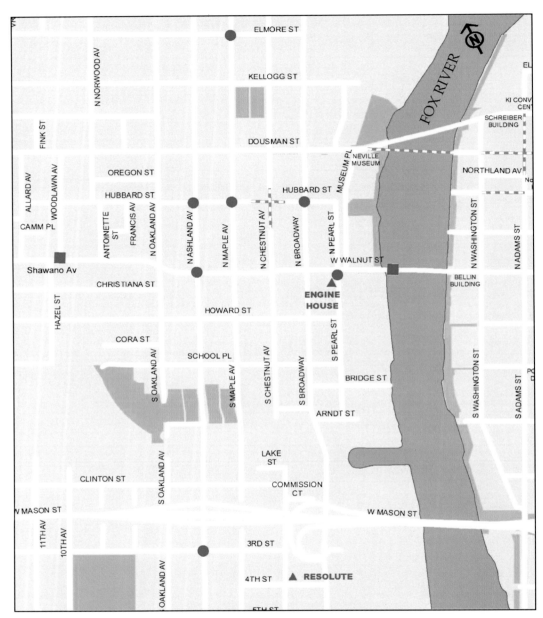

Fort Howard Fire Department-related locations (1873-1895) within modern Green Bay.
The fire company houses are indicated by the triangles. Water supply sources are indicated by the circles for cisterns and the squares on the bridges for access holes. The bridge access on Shawano was over a slough (tidal creek) that since has been filled in. Streets that no longer exist as of 2016 are indicated by dashed lines.

Chapter 10

Gradual Changes
The Fort Howard Fire Department
1873 - 1895

The Fort Howard Fire Department and borough were both formally founded in 1856. Starting from nothing, the Fort Howard Fire Department kept pace with the changing community. The community grew and developed, becoming a city in 1872. From that point until annexation by Green Bay in 1895, changes to the fire department were inevitable and often reflected advancing expectations from the community.

Fort Howard Fire Department was fairly active through the 1870s. Several bills for hauling the steamer were often presented at most Common Council meetings. Coal, used as fuel for the steamer, was a recurrent cost. Fires were frequent.

The numerous fires prompted changes in the fire

Changes to the fire department were inevitable and often reflected advancing expectations from the community.

> *The Common Council paid $2 bounties to the first firefighter to "fire the steamer."*

department. For instance, the Common Council paid $2 bounties to the first firefighter to "fire the steamer." Although the steamer was connected to a water heater at the engine house, a fire in the burn box was still necessary to fully heat the water to a boil. The sooner the fire was started in the burn box, the quicker steam was produced, and the faster water streams could flow. "Fireing the steamer," as it was called, meant to start a fire in the burn box using paraffin- or kerosene-soaked rags and kindling. This produced the belching, dark smoke, often seen in images of steamers. The steamer was fired as soon as possible after the fire alarm sounded.

The engineer of the steamer had foremost responsibility to fire the steamer. If he was unavailable, the first firefighter to reach the engine house after an alarm sounded handled that duty. The first instance of a bill for fireing the steamer was in November 1875. The practice continued until the 1895 annexation.[1]

The operating routines of the Fort Howard Fire Department were becoming more advanced. The increased number of fire calls in the 1870s revealed an issue that took many years to solve. Horse teams hauled the steamer to the fire, but they were first brought to the engine house in response to the fire alarm bell. These ordinary citizens

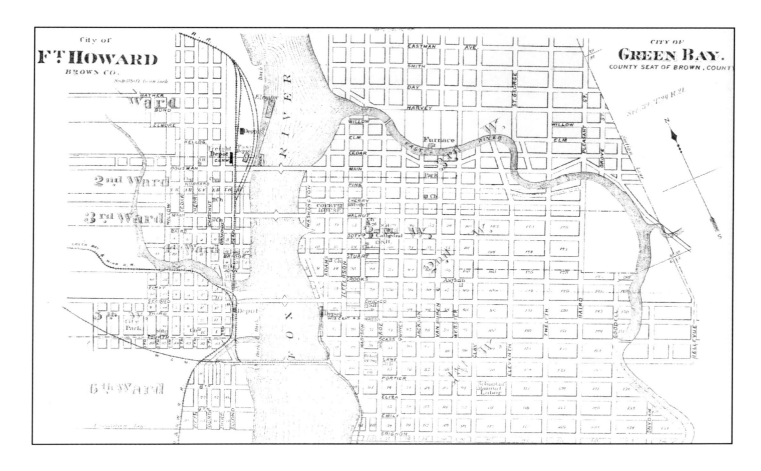

Map of the cities of Fort Howard and Green Bay from 1881.

or firefighters were paid a bounty as an incentive for their volunteer service. Similarly, horse teams hauled the hose carts, with an incentive bounty also paid to the owners. Under this arrangement, response times of the steamer and hose cart to the fire scene were entirely dependent on the timely arrival of horse teams at the engine house. If teams did not respond, the steamer and hose carts were pulled by the firefighters themselves, which was a slow and exhausting process.

The voluntary horse team system was only somewhat reliable, especially at night when the owners wished to sleep.

At an August 21, 1876, meeting, the Fire Department Committee acknowledged that while the steamer would arrive first, the hose cart was often excessively delayed, especially in the case of wet weather when the dirt streets became muddy quagmires. The hose cart was absolutely necessary for firefighting operations. Firefighters deployed the fire attack hose (stored on the hose cart) from the steamer to the building on fire.

The committee proposed the purchase of a four-wheeled hose cart, which would be better suited for dirt streets.[2] The July 27, 1876, *Fort Howard Herald* supported this proposal, stating that pulling the two-wheeled hose cart exhausts the firefighters. In late 1876, Fort Howard purchased a four-wheeled hose cart, able to hold 1,000 feet of hose, for $700.[3] Although better able to negotiate dirt streets, the size and weight were substantial. Thus, horses were still the best option.

The voluntary horse team system was only somewhat reliable, especially at night when the owners wished to sleep. At the chief engineer's suggestion, the Common Council contracted with owners of horse teams in February 1877 to pull the steamer and hose cart from 6 pm to 7 am. Other owners were not compensated unless the designated teams were more than five minutes late.[4] This system was an improvement, but not without lapses.

Eventually, the city considered purchasing its own horses for the fire department. In July 1883, a proposal was made to purchase "a team of horses to be used for hauling the Steamer, and working on the Streets, etc. Team to be kept in the barn near the Engine House, teamster to be hired by the City."[5] Under this proposal, a team would be dedicated to the fire department, but used for other city purposes when not engaged at a fire. This proposal was not consummated because of costs. However, one month later, the Common Council accepted a contract to keep a horse team at a barn near the engine house from 7 pm to 6 am for $25 per month.[6] While similar to the previous arrangement, this contract kept the horses near the engine house at night.

This arrangement met with limited success over the next year. Eventually, taxpayers complained about fire department inefficiency because "no team or teams could be had to haul the Steamer or hose cart."[7] In response, the city purchased two horses in February 1885 for $390.[8]

Just as proposed nearly three years before, these horses stayed in a barn by the engine house at night and

"No team or teams could be had to haul the Steamer or hose cart."
- Fort Howard Common Council minutes

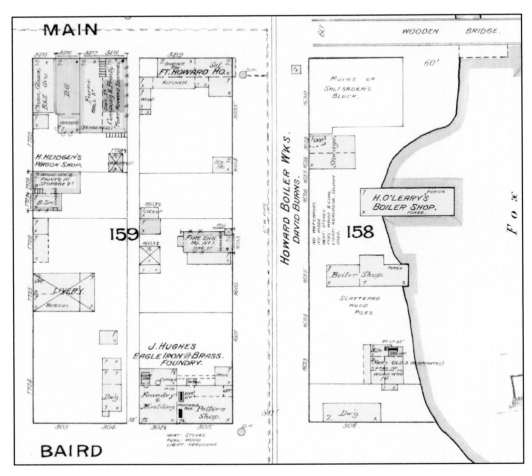

Fort Howard Fire Department Engine House on South Pearl Street from the 1887 Sanborn Insurance map (left-center). Main Street to the north is modern West Walnut, while Baird is now Howard Street. The riverbank has been substantially extended outward through the decades. A closeup of the engine house location is on the next page.

performed city-related work on the streets during the day. In the event of a fire alarm, the other work stopped and the horses immediately hauled the steamer. In addition, the city hired one horseman (known as a teamster) to work the horses.[9] The teamster had the additional duty of keeping holes open through the ice on the Fox River, which were used as water supply access points for the steamer.[10] Fort Howard Fire Department logistics had improved.

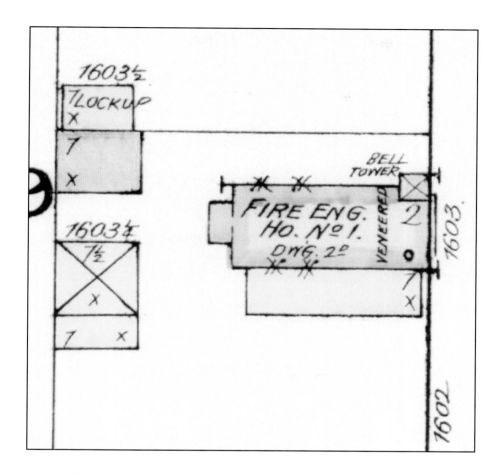

Closeup of the Fort Howard Fire Department Engine House on South Pearl Street from the 1887 Sanborn Insurance map (full image on the previous page). This engine house opened in February 1873 and served as Station No. 3 until 1937.

Beyond horses and equipment, Fort Howard addressed the organization of the firefighters. At an April 1883 Common Council meeting, a fire department statement complained that in spite of a population of about 3,000, they had "great difficulty in filling up their ranks with volunteers necessary for a first class organization."[11] An insurance yearbook from 1882/1883 tallied twenty-two volunteer Fort Howard firefighters, which must have been inadequate.[12]

In response, the Common Council authorized the Fort Howard Fire Department to consist of fifteen firefighters maximum, but now each was paid $12 per year. For each fire a member missed, they deducted $1 from his pay, which was then divided amongst those firemen who did respond. In addition, the chief engineer received $30 per year.[13] With these arrangements, Fort Howard Fire Department became a part-time agency in 1883. As part of an annual fire department review two years later, the fire company secretary reported that some members "absented themselves from every fire." Consequently, the Common Council reduced the authorized number of firefighters to ten, but increased yearly pay to $15.[14] The Fort Howard firefighters had become an elite, dedicated force.

Fort Howard continued to improve its firefighting water supply. When first founded, most of the borough was close enough to the Fox River that it was a suitable water source. Two below-ground cisterns built in 1869 were located in newly developed neighborhoods. These were several blocks from the river, too far for it to serve as a water source.

As Fort Howard continued to grow, so did its need for water supply. Additionally, the steamer purchased in 1872 had the capacity to flow much more water than the hand pumper. Consequently, the Common Council approved construction of another cistern, supplied from the river by an underground pipe, at North Broadway and Hubbard Street in 1874 to help satisfy the large volume of water needed at fires.[15] One year later, Fort Howard built a fourth cistern near the South Pearl Street Engine House.[16]

Going even further, Fort Howard Mayor George Richardson stated during his April 11, 1878, inaugural address that the Fire Department Committee needed to establish more water reservoirs as the steamer was "comparatively useless" in a "large proportion of the city."[17] This is borne out by the complete loss of a dwelling about a year before in the southern portion of Fort Howard. The fire department responded promptly, but the "fire was too far from water for them to be of service."[18] Within a few months of the inaugural promise, the city built two additional reservoirs in the expanding southern part of the city.[19]

Also in 1878, officials placed a fire department access box on the Wolf River Road Bridge over the slough, which today would be on Shawano Avenue just east of current Green Bay Fire Department Station 3.[20] The slough was essentially a tidal creek that passed through what

The steamer is "comparatively useless" in a "large proportion of the city."

- Fort Howard Mayor George Richardson

City of Fort Howard from the 1893 lithograph. Industry thrived along the riverbank and Pearl Street (closest to the river), while mercantile occupied the Main Street (modern West Walnut) and Broadway areas. (See a closeup of the highlighted section on the next page and the full-size lithograph on page 180.)

was then the west side of Fort Howard and emptied into the Fox River near Mason Street. The slough was filled in over the decades, but remains evident today in the low area susceptible to flooding along Ashland Avenue at Seymour Park and Shawano Avenue in the area around Station 3.

Firefighters parked the steamer on the bridge. They passed the supply hose through the box into the relatively shallow slough, which still provided ample water supply. This bridge marked the far west edge of the expanding city. A reservoir was built at what is now Ashland and Third Street in August 1881, and in July 1883, two cisterns were built at the intersections of North Maple/Elmore and North Maple/Hubbard streets.[21] Overall, five

Fort Howard Fire Department Engine House from the 1893 lithograph. The engine house is amongst industry and businesses. (From highlighted section of the image on the previous page. See the full-size lithograph on page 180.)

> *While the Fort Howard Fire Department advanced in the aspect of water supply, it regressed in terms of staffing.*

cisterns and an access box on the slough bridge were built from 1875 to 1883, greatly improving firefighting water supply in the expanding city.

There were no major changes to the Fort Howard Fire Department for the next several years. The city continued to keep a horse team and teamster performing street work during the day, but available at any time to pull the steamer.[22] This seems to have been effective, as the number of additional bills for horse teams hauling the steamer dropped substantially. The Common Council routinely paid the ten authorized firefighters a total of $150.[23] The fire department continued the full-time engineer of the steamer position, although they reduced the salary to $500 per year in 1877.[24]

As in Green Bay, 1887 was a significant year of change for the Fort Howard Fire Department. The Green Bay and Fort Howard Water Works Company began operations, providing pressurized water through underground mains connected to fire hydrants. The system passed an acceptance test in June 1887.[25] After relying on river access and cisterns, this was a dramatic improvement to water supply for fighting fires.

The city paid the Water Works $3,000 per year

hydrant rental.[26] To offset this expense, the city discharged the Engineer of the Steamer one month after the acceptance test.[27] The belief that the steamer fire engine itself had become unnecessary played a role in the decision to eliminate the engineer. Water volume and pressure from the hydrants was great enough that highly effective water streams came from hoses attached directly to the hydrants. This configuration completely bypassed the steamer pump.[28] While the Fort Howard Fire Department advanced in the aspect of water supply, it regressed in terms of staffing.

The capacity of the Water Works made the steamer seem obsolete. Municipal authorities, and perhaps some firefighters, thought the steamer had become unnecessary. In August 1887, two months after the successful acceptance test of the water supply system, the Fort Howard Common Council sought to "lay up [the] engine in proper manner;" that is, take the steamer fire engine out of service.[29] This was not actually done. The Common Council discussed the proposal a few months later, but again took no action.[30]

Fortunately, the short-sighted belief that the fire department no longer needed the steamer did not prevail. Bills for hauling and fireing the steamer continued to be presented to the Common Council through the next few years. The fire engine was still utilized, albeit not as often as before the Water Works.[31] Firefighters used the steamer at fires distant from hydrants or to utilize the river at larger fires as a supplement to hydrant supply.[32] As with so many firefighting tools that are not used often, the value of the steamer was truly revealed only when called upon.

The fire department expanded in the fall of 1887 to better protect the expanding south side of the city. Back in 1872, a lot on the north side of Fourth Street, between Chestnut and Broadway, was purchased as a site for a fire alarm. First there was a simple triangle, and then a bell.[33] In fact, an 1875 tax record refers to this location as "Fire Alarm Bell."[34] In September 1887, three months after the Water Works acceptance test, the fire department placed a hose cart and hose in a simple shed with the fire bell tower at the Fourth Street location.[35] The shed and tower were improved that winter to protect the hose from freezing, and completely rebuilt the following summer.[36] Shortly thereafter, Fort Howard Resolute Fire Company was formed and based at the new south side hose cart shed.[37] The Common Council provided this new company with rubber coats and new hose so that by 1892, the company had 850 feet of good hose and a used hose cart from Green

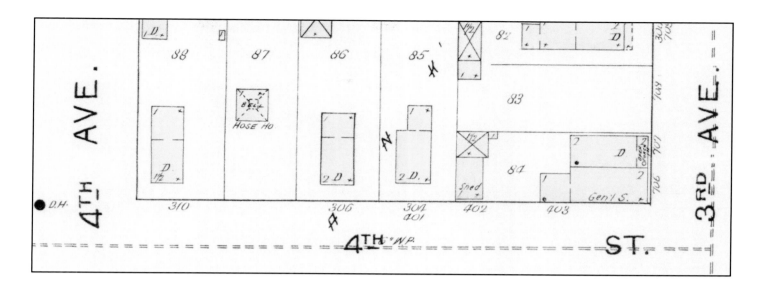

Resolute Fire Company hose cart shed from the 1894 Sanborn Insurance map (left-center under the '87'). This site is on the north side of 4th Street between modern South Broadway (3rd Avenue) and South Chestnut (4th Avenue) and is now residential parking. As early as 1872, a fire alarm bell was on this site. In 1887, the City of Fort Howard built a shed and placed a hose cart there. One year later, the newly created Resolute Fire Company began to respond with this hose cart.

Bay.[38] The Resolute Company members were drawn from the ten existing, authorized part-time firefighter positions. In a January 1890 report, the chief engineer stated there were still ten total Fort Howard firefighters.[39]

Creation of the Resolute Company was an unquestionable result of the capacities of the new Water Works. Upon a fire alarm, the Water Works increased water pressure within the entire water system, on both sides of the river, to 100 pounds per square inch (psi).[40] This pressure gave very good water streams from hoses attached directly to hydrants. A fire engine was no longer even used. The Resolute Company firefighters attached hose to a hydrant, deployed this from the cart, then opened the hydrant and initiated a fire attack without any involvement

> *On two occasions, the Water Works staff did not hear the fire alarm bells in Fort Howard because of "adverse winds" and the distance across the Fox River.*

of the steamer. This fire company needed only hose, cart, and shed, completely avoiding the tremendous expense of a fire engine.

Resolute Company was fairly active, responding to six fires in 1888, its first full year of existence.[41] Fire protection in the south side of the city was vastly improved, no longer relying on equipment responding from the engine house in the center of town.

A significant, controversial event occurred in 1887. The assistant chief, an alderman, and the city marshal conspired to demonstrate the improved response time and capabilities of the expanded Fort Howard Fire Department. On October 22, 1887, they tried to start a bonfire in the south side of the city to initiate a fire department response. However, conditions were too wet, so the assistant chief simply rang the bell at the Resolute Company hose cart shed. This triggered a full fire department response to what was a deliberate false alarm. This inappropriate, essentially illegal act and conspiracy, were quickly exposed.

The Fort Howard Common Council took as series of depositions, revealing the complicity of the conspirators.[42] Chief Engineer A. L. Gray testified that nobody is authorized to ring the fire bells except in the case of fire.

Regardless, the Common Council minutes make no mention of any consequences. In fact, the *Daily State Gazette* reported that after accepting a summary of the depositions, the Common Council took "no further action in the matter."[43] It appears the matter was dropped at this point.

Pressurized water supply from hydrants dramatically improved firefighting capabilities. However, a problem surfaced for firefighting efforts in Fort Howard. The Green Bay and Fort Howard Water Works Company provided a constant domestic pressure of 40 psi. Upon notification of a fire alarm, additional pumps were utilized to provide system-wide "fire pressure" of 100 psi.[44] As part of his January 1892 annual report, the Fort Howard Chief Engineer noted there were eighteen fires and false alarms during the previous year. On two occasions, the Water Works staff did not hear the fire alarms bells in Fort Howard because of "adverse winds" and the distance across the Fox River. As a result, the Fort Howard firefighters had to contend with 40 psi domestic pressure, which was completely inadequate for firefighting.[45]

The Water Works facility was located in the same location as the modern Green Bay Water Utility site: on South Adams Street just south of East Mason on the east side of the Fox River. Water Works employees on the

Green Bay side of the river had to hear the fire alarm bell ringing in Fort Howard to recognize the need to increase to fire pressure. This system was unreliable. Consequently, in the fall of 1892, an electronic fire alert mechanism connected the South Pearl Street engine house in Fort Howard and the Water Works facility on South Adams in Green Bay.[46] This addressed the chief engineer's concern that this situation could "prove disastrous to our city."[47]

Another concern was the vulnerability to failure of the large water mains crossing the Fox River from the primary pump station in Green Bay. The Water Works Company built a pump station with a new artesian well in Fort Howard in 1894, fully resolving this issue.[48] From then on, the Fort Howard Common Council never again dealt with significant water system problems. Firefighting water supply in Fort Howard was sufficient and reliable.

Fort Howard and its fire department grew in tandem through the early 1890s. Citizens in new neighbor-

City of Fort Howard in 1891. Taken from Green Bay, this image shows the bridge between Walnut in Green Bay and Main Street (later West Walnut) in Fort Howard. Industry and businesses stretch all along the riverbank. The Fort Howard fire station at the middle left is mostly obscured.

hoods requested water main extensions and additional hydrants. When operations began in 1887, there were six miles of mains and seventy-four hydrants in Fort Howard. By 1895, there were nine miles of mains and ninety-four hydrants.[49] Next, on January 5, 1894, the annual fire department report noted there had been twenty-three fire calls in 1893, a "greater number than any previous year."[50] The city horse team continued to perform municipal streets task, but was always available to haul the steamer.

The Common Council eventually ordered the work limited to within a reasonable distance from the engine house and never on country roads.[51] To further reduce response times, the fire department installed "drop harnesses" in the engine house in 1894 to quickly hitch the horse team to the steamer.[52] The Fort Howard Fire Department was becoming more sophisticated.

In contrast to the Green Bay Fire Department, changes to the Fort Howard Fire Department were not incident-driven, at least not directly. Major fires led to significant changes in Green Bay. Fort Howard experienced those same changes, but not because of fires on the west side of the Fox River. Changes in Green Bay were reactive, while the same changes to the Fort Howard Fire Department appear to be proactive, almost mundane—unless one considers the events in Green Bay as influences on Fort Howard.

Although never stated in the Fort Howard newspapers or Common Council minutes, it is reasonable to conclude the lessons learned in Green Bay were applied equally in Fort Howard. For example, when the Borough of Fort Howard was first established in 1856, a fire company was formed and supported right away. Consequently, the Fort Howard Fire Department was immediately permanent, unlike the three failed efforts that occurred in Green Bay from 1842 to 1851.

Fort Howard saw the benefit of the Green Bay hand pumpers, and therefore immediately rejected an unreliable used engine and purchased a new model. Similarly, just four years after Germania No. 1 obtained a steamer fire engine in Green Bay, Fort Howard purchased its own steamer and similarly hired a full-time engineer of the steamer. Fort Howard built river access points and cisterns shortly after Green Bay to address water supply issues. Lastly, while the Water Works was initiated in Green Bay as a consequence of the 1880 conflagration, Fort Howard simply joined in to create a common water supply system.

Essentially, there were no precedent-setting fires in Fort Howard. Changes on the west side of the river mirrored changes on the east side in Green Bay. However, in 1895, these parallel, time-offset changes came together, bringing a tremendous transformation to Fort Howard, the fire department, and that of its cross-river sister city, Green Bay.

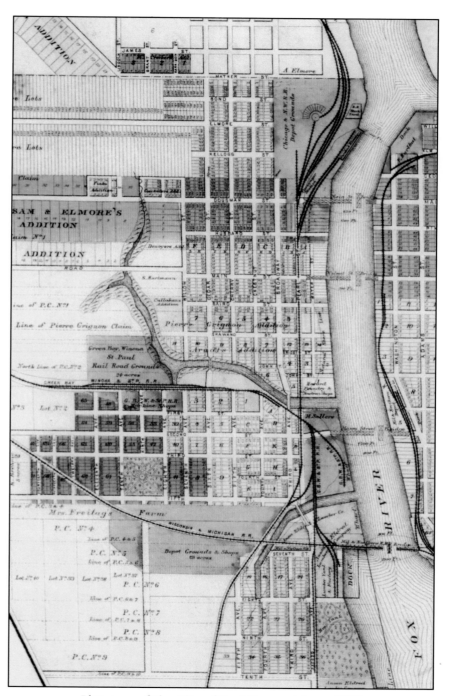

Plat map of the City of Fort Howard from 1890

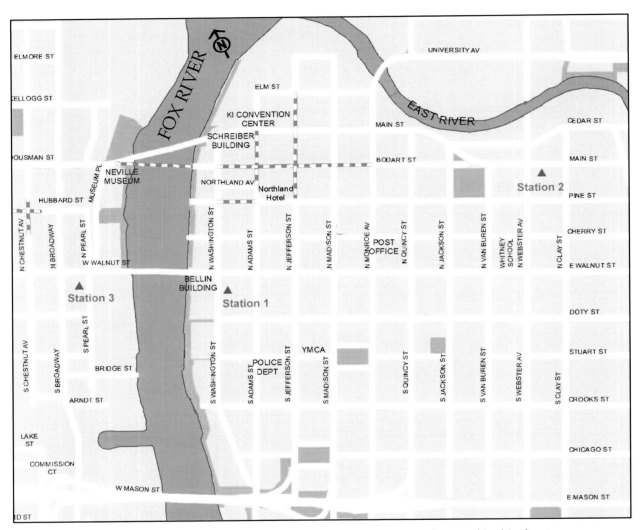

GBFD fire stations locations (1895) within modern Green Bay. Fire stations discussed in this chapter are indicated by the triangles and labeled with the station designation. These stations remained open until replaced in 1929 for Station 1, 1965 for Station 2, and 1937 for Station 3. Streets that no longer exist as of 2016 are indicated by dashed lines.

Chapter 11

Mutual Aid Leads to a Merger
1895

It was always an obvious possibility, if not an inevitability, that the cross-river neighbors of Green Bay and Fort Howard would merge into one community. Separated by only a few hundred feet of the Fox River, through time the communities had become interwoven at many levels.

The first US soldiers at the Fort Howard garrison reported friendly relations with the Green Bay inhabitants. As early as 1842, Charles Wheelock operated a ferry across the Fox River, facilitating improved social and commercial interactions.[1] The first railroad in the area came to Fort Howard in 1862.[2] This prompted construction of the first bridge across the Fox the next year so Green Bay could take full advantage of the railroad.[3] Two more bridges followed in 1874.[4] Each bridge enhanced the connection between the cities, inspiring one newspaper editorial to

The fire departments of Fort Howard and Green Bay supported each other early, often, and enthusiastically.

*Mutual aid -
The sharing of firefighting resources between communities.*

describe the "mutual advantage" to both cities.[5] Although commerce was somewhat diverse, the two cities developed similarly and in parallel.

The Green Bay Common Council first considered annexing the three-year-old Borough of Fort Howard in 1859. They sent a bill to the state, but it was not acted upon.[6] In 1872, Fort Howard voters resoundingly rejected annexation with 27 ballots for and 355 against.[7] Two years later, the Green Bay Common Council proposed annexation, but their Fort Howard counterparts declined this offer too.[8] At an 1889 open meeting, the citizens of Fort Howard refused yet another offer of annexation from Green Bay.[9] One newspaper observed the successful annexation effort in 1895 was "nothing new, having been several times agitated during more than twenty years."[10]

For nearly three decades, the municipal governments remained separate. In contrast, the fire departments of Fort Howard and Green Bay supported each other early, often, and enthusiastically. Sharing firefighting resources between communities is known today as mutual aid. This practice thrived between the young communities of Green Bay and Fort Howard. While the municipalities remained distinct, the division between the fire departments was less so, especially at the operational level.

The first instance of mutual aid came in August 1855. A wood mill burned on the west side of the river in not-yet incorporated Fort Howard. A sixteen-year-old worker was killed. Fort Howard citizens fought the fire—they did not yet have an organized fire company or even a fire engine. The newly organized Germania No. 1 fire company from Green Bay also fought the fire. These Green Bay firefighters operated the brakes of their hand pumper for about eight hours, protected adjacent property, and earned great praise from the mill owner.[11] Though not reported, the hand pumper fire engine could only have been brought to the west side by boat. There were no bridges across the river at that time. The mill owner, T. O. Howe, made a substantial financial donation to GBFD in a demonstration of appreciation four years later.[12] Time had not diminished his gratitude.

Acts of mutual aid continued. In August 1856, a barn at the Blesch Brewery on North Broadway caught fire, and the blaze spread to adjacent buildings.[13] Fort Howard still did not have a fire department. Once more, Germania No. 1 crossed the river to help. Charles Wheelock promptly brought the hand pumper and fire-

A CARD.
The undersigned would take this method of returning their thanks to the Fire Departments of Green Bay and Fort Howard respectively, for the arduous and highly efficient service rendered during the late fire, and which was the means of saving our establishment from the flames.
TAYLOR & YOUNG.

Newspaper notices from the January 25, 1868 (above), and August 8, 1868, *Green Bay Gazette*. These fires occurred in Fort Howard, and the grateful fire victims explicitly thank the fire departments from both cities. Cooperation between the cross-river fire departments occurred often and willingly.

A CARD.
The undersigned hereby return their sincere thanks to the citizens and Fire Departments of Green Bay and Fort Howard, and especially the members of Guardian Engine Co., No. 2, for their services in preventing their mill from burning on Wednesday evening last.
E. C. BAIRD, & Co.
Green Bay, Aug. 7th, 1868.

Mutual Aid Leads to a Merger

> *The ease and speed of mutual aid responses across the first newly opened bridge now "makes the fire departments of the two places, one, which is no small advantage to either side."*
>
> — *Green Bay Advocate*

fighters across the river on his ferry. Wheelock and the Germania No. 1 firefighters again received great public praise from Fort Howard citizens and municipal officials. A few months later, Wheelock became a charter member of the newly formed Guardian No. 2 fire company in Green Bay.[14]

In July 1859, the largest fire to date in Fort Howard destroyed a business and dwelling. The fire was so large that the two Green Bay hand pumpers and one hook-and-ladder truck were brought over by ferry and worked alongside the Fort Howard hand pumper and firefighters.[15] The Green Bay Common Council paid the 25-cent ferry toll.[16]

Mutual aid occurred in both directions. In 1860, Fort Howard Fire Department fought a fire on Washington Street in downtown Green Bay. The newspaper described how the "four excellent engines [three Green Bay and one Fort Howard] threw a perfect torrent of water," thereby limiting the damage.[17] Once, in 1862, Fort Howard unsuccessfully attempted to cross to Green Bay for a fire. The ferry was on the Green Bay side of the river and could not be summoned for the 5:30 am fire alarm.[18]

The harsh Northeast Wisconsin weather did not prevent mutual aid. In December 1861, Green Bay firefighters carried their hose, hooks, and ladders across the ice-covered Fox River. The ice was not thick enough to safely cross with carriages.[19] This great act of commitment demonstrated the zeal with which firefighters were willing to assist. The January 31, 1861, *Green Bay Advocate* complimented the Fort Howard firefighters for responding to Green Bay when snow made travel difficult. Going further, the newspaper predicted annexation and that the Fort Howard Fire Department will "be 'one of us' in organization as they are now in practice."

There were fires in both cities over the course of a few hours on the same day in June 1864. Mutual aid was requested both times. Consequently, a newspaper editorial suggested the ease and speed of mutual aid responses across the first newly opened bridge now "makes the fire departments of the two places, one, which is no small advantage to either side."[20] The communities considered the fire departments so closely allied that the operational separation was becoming blurred.

Green Bay officials recognized that formal rules were needed. An 1858 ordinance detailed the first rules for the Green Bay Fire Department (GBFD) and included provisions on mutual aid. Specifically, fire companies could go outside Green Bay city limits only if granted permission by the chief engineer (modern term is fire chief).

The mayor or president of the Common Council had to give consent as well. If a fire company left the city limits without permission, the officer faced a $10-$25 fine.[21]

It is unknown if any officers faced punishment. The ordinance did not specifically name Fort Howard as a destination. However, in 1858, this was the only substantial community near Green Bay, and therefore, the most likely recipient of mutual aid.

GBFD Hose Company No. 3 in the late 1890s. The face pieces on the horses' foreheads are adorned with the number "3" signifying the fire company. The driver is John Walters, one of the first four west side GBFD firefighters hired in May 1895. The other three are unidentified. Notice the dog sitting with the driver.

The two fire departments interacted beyond fighting fires. Fort Howard firefighters participated in the July 4, 1859, parade in Green Bay.[22] Two months later, Fort Howard Fire Department received a new hand pumper. The ferry brought it over to Green Bay for a "wetting down" celebration. The event included a competition against the new Green Bay Guardian No. 2 hand pumper.[23] When GBFD Washington Hook and Ladder Company No. 1 formed in January 1858, they offered "honorary" membership to residents of Fort Howard.[24]

The 1860 Brown County Fair included a formal fire

department competition between the three Green Bay and one Fort Howard hand pumper companies. Green Bay Guardian No. 2 generated the longest stream and won an ornate trumpet (see page 46) and bragging rights. The paper described the event as "characterized with greater harmony than usual on such occasions; the little differences arising amounting merely to a war of words."[25]

The 1869 annual GBFD review included the Fort Howard engine.[26] In 1870, the Fort Howard hand pumper, hose cart, and firefighters joined in the July 4th parade in Green Bay.[27] When the Fort Howard steamer arrived in March 1872, firefighters held a side-by-side trial with GBFD Enterprise No. 1 steamer.[28] Overall, it seems relations between the two fire departments were amicable.

Green Bay Chief Engineer Samuel Lindley announced a mutual aid alerting system in 1872. After the bells at the GBFD stations sounded a general alarm, additional bell taps indicated the general location of the fire. One tap was for Green Bay First Ward, two taps for the Second Ward, and three taps for the Third Ward.

Hose Company No. 3 driver John Walters with one of the company horses.

Mutual Aid Leads to a Merger

> *This is believed to be the first formal mutal aid arrangement between the fire companies.*

Most significantly, four taps indicated a response into Fort Howard.[29] These signals were repeated to ensure location recognition. This is believed to be the first formal mutual aid arrangement between the fire departments.

Fort Howard also used the bell at its engine house to summon mutual aid.[30] In fact, an ordinance forbid anyone but the chief engineer, assistant, engineer of the steamer, mayor, or an alderman from ringing the bell to summon mutual aid.[31] In some instances, the bell sound was too faint to be heard across the river.[32] This led to delayed mutual aid responses from Green Bay.

By 1872, there were two steamer fire engines with GBFD and one with Fort Howard. There were three fires in Green Bay on consecutive days at the end of July, and Fort Howard assisted at each. In one case, broken carriages prevented GBFD Enterprise No. 1 steamer and hose cart from responding. The newspaper reported that the Fort Howard firefighters were even more valuable than usual in this circumstance.[33]

In October that year, all three steamers operated at a fire in Fort Howard, most likely together for the first time.[34] Bystanders and firefighters must have been impressed by the volume of water flowed. This large fire destroyed a stretch of businesses near the Walnut Street Bridge. The damage would have been worse but for the suppression efforts supported by three steamers.

The positive state of the mutual aid relationships helped avert problems in both communities. Fires occurred in Green Bay and Fort Howard on separate days in June 1872. In each case, the steamer from the other city "had first stream on," meaning the cross-river neighbor first flowed water on the fire, beating the home-town steamer. In spite of potential embarrassment, both the newspapers and municipal officials showed appreciation.[35] Relations were so positive a *City Directory* passage stated:

> "It is a matter of credit to the different fire companies of Green Bay and Fort Howard that they mutually assist each other and in this manner increase the efficiency of each. Such good feeling and unity of purpose cannot but prove pleasant to their members as also beneficial to the cities."[36]

Two unique situations further reveal the high level of support. First, Fort Howard sold its only hand pumper in April 1871.[37] The new steamer fire engine arrived in March 1872, nearly a year later.[38] Fort Howard Fire Department was without a fire engine during this time. Although not revealed in any records, it's reasonable to conclude that Fort Howard relied on GBFD for fire en-

The positive state of the mutal aid relationships helped avert problems in both communities.

gines. The second instance occurred after fire destroyed the Fort Howard engine house on November 6, 1872.[39] The steamer was saved, with only minor damage.[40] Instead of using temporary quarters on the west side, GBFD housed the steamer at Engine House No. 1 on South Washington Street until the new Fort Howard station opened a few months later.[41]

Steamer fire engines arrived in Green Bay in 1868 and 1872, and in Fort Howard in 1872. It was best to haul these heavy apparatuses by horse. Neither fire department had horses, so the cities paid a bounty to horse team owners to haul the steamers. This included mutual aid responses.

However, the cities did not have a formal horse bounty agreement during the first two decades of mutual aid. While there were many instances of mutual aid, in only a few instances are payments recorded in official records. For example, the Green Bay Common Council paid $10 to Phillip Franck for hauling the GBFD steamer to a fire in Fort Howard in 1864.[42] Fort Howard paid horse team owners for bringing a GBFD steamer to fires twice in 1869.[43] Fort Howard paid for the GBFD steamer in 1870, while in 1872, Green Bay paid expenses for the Fort Howard steamer.[44] It seems officials had not established a payment arrangement and practices were inconsistent.

Eventually, lack of a formal mutual aid agreement regarding horse team bounty payments led to controversy and conflict. In May 1874, GBFD responded into Fort Howard for what was ultimately a false alarm.[45] GBFD Chief Engineer Louis Scheller submitted a horse team bounty bill to the Fort Howard Common Council. In a surprise move, the Fort Howard aldermen refused to pay the bill, ostensibly citing that the letter must come from an "official" such as the mayor or an alderman.[46] The Green Bay Common Council quickly confirmed that the chief engineer was indeed authorized.[47] Fort Howard municipal officials did not reply. The disagreement continued and an impasse developed.

In an effort to resolve the issue, Scheller proposed "an arrangement for the use of the fire extinguishing apparatus of the respective cities for mutual benefit."[48] In particular, he recommended that the city having the fire would pay expenses.[49] The Fort Howard Common Council

Interior of a fire station in the late 1890s or early 1900s. Though there are no markings, this image is from the collection of GBFD Fire Chief William Gleason, so it may be a GBFD station. These upstairs living quarters are rather spartan. Notice the brass pole to the apparatus floor and the four beds.

Mutual Aid Leads to a Merger

"refused to consider the proposition at all."[50] The Green Bay aldermen felt the proposal was fair and expressed surprise that the Fort Howard Common Council "has resolved to treat [the proposal] with contempt."[51] In response, Scheller confirmed that GBFD companies would not leave the city unless directly ordered.[52] GBFD firefighters supported their chief by announcing they would strictly abide by the ordinance and not leave Green Bay city limits without orders.[53] The impasse had deepened.

The deadlock resulted in significant consequences within two weeks. Fort Howard Fire Department responded to a fire in their city on modern Walnut at Pearl Street, near the bridge. Unfortunately, the steamer malfunctioned and water could not be sprayed. Consequently, the small fire in an upper story quickly spread.[54]

The two GBFD steamer fire engines responded, but stopped on the Green Bay side of the bridge, waiting for official orders, which never came. After a long delay, Fort Howard citizens directly and urgently implored GBFD to help. Furthermore, A. Weston Kimball, an insurance agent, promised to pay expenses.[55] The Green Bay firefighters crossed the bridge and extinguished the fire, but not before the blaze destroyed two buildings and damaged another.

> *The two GBFD steamer fire engines responded, but stopped on the Green Bay side of the bridge, waiting for official orders, which never came.*

The *Green Bay Advocate* asserted the fire grew and extended greatly during the delay. Pointing out the folly of the situation, they estimated "the consequent loss by the one fire is equal to what it would probably have cost Fort Howard for hauling our engines for twenty years."[56] The newspaper also pointed out that with four times the population, Green Bay had four times the number of fires, arguing the cost is "a wash" and expense should not be a consideration for fire protection.[57]

Citizens, firefighters, and the newspaper staff recognized the reality of the situation, but municipal officials did not. In August 1874, Fort Howard Fire Department assisted at a barn fire in Green Bay at Pine and Quincy.[58] The very next day, the Fort Howard mayor wrote a letter to that city's chief engineer, which was reprinted in the *Fort Howard Herald*. He ordered that the fire bells in that city were to be rung only for fires there "and no other," and Fort Howard steam fire engine was not to leave the city limits.[59] This apparently retaliatory response effectively ended mutual aid.

The *Green Bay Advocate* called attention to the absurdity of the situation. The newspaper reported that in the two weeks since the Fort Howard mayor ordered mutual aid halted, there had been six fires in Green Bay, but

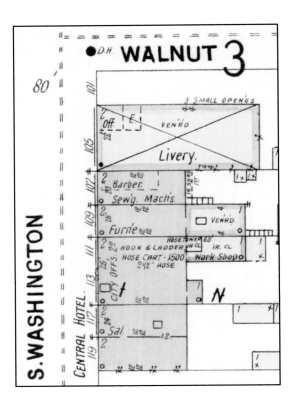

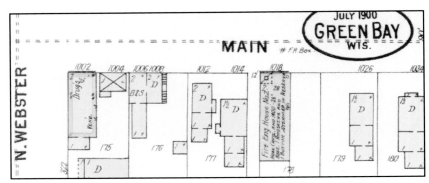

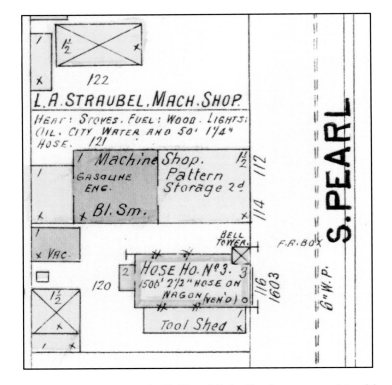

The three GBFD fire stations from the 1900 Sanborn Insurance map. Following the 1895 merger, GBFD consisted of a hook-and-ladder company and hose cart company at Station 1 on South Washington (above); hose cart company and two reserve steamers at Station 2 on Main Street (top right); and a hose cart company at Station 3 on Pearl Street (right).

none in Fort Howard. If Chief Scheller's proposal had been accepted, Green Bay would have paid $72 in expenses, whereas Fort Howard would have had none.[60] Municipal officials remained unmoved and the mutual aid standstill continued.

The citizens continued to suffer from fire because of the political impasse. In early 1875, a fire occurred on the south side of Fort Howard. The Fort Howard steamer arrived about an hour after the alarm, having been drawn by hand because horses were not available. GBFD steamers "refused" to respond because of the "imbroglio existing between the fire departments of both cities."[61] The *State Gazette* reported a GBFD steamer "was applied for, but owing to the unhappy relations on fire matters existing between the two places, a delay finally rendered it useless to send it to the rescue."[62] The fire destroyed a store and two homes. The newspaper absolved the GBFD engineers of any responsibility and squarely placed blame on the politicians from both cities. The paper argued, "It should not be necessary to obtain the Mayor's or anybody else's authorization to take the engines to Fort Howard in case of fire."[63]

The politicians finally relented after the community endured a second fire made worse as a consequence of the dispute. Officials negotiated a formal mutual aid between Green Bay and Fort Howard fire departments. The mayors and chief engineers from both cities signed the agreement at the end of January 1875.

Specific terms were set out. Officials requested

> *"It should not be necessary to obtain the Mayor's or anybody else's authorization to take the engines to Fort Howard in case of fire."*
>
> *- State Gazette*

Mutual Aid Leads to a Merger

mutual aid by four continuous taps of the fire bell, as established years earlier. A steamer would respond to fires within three blocks of the river. The two Green Bay steamers alternated months to be "on duty" for mutual aid. Fort Howard could request the second GBFD steamer. Most importantly, the providing department paid the horse bounty cost. If the second GBFD steamer was requested, Fort Howard paid the additional hauling cost.

The *Fort Howard Herald* stated, "An amicable agreement has been arrived at between Green Bay Fire Department and that of this city. We can see no cause for disagreement. Let brethren dwell together in unity."[64] Fort Howard Mayor Richardson hoped, "That the fullest of friendship and fraternity be encouraged in the fire departments of the two cities; that all feelings of jealousy and enmity be laid aside, and discountenanced in the future."[65]

The new agreement succeeded. Within a few months, GBFD Guardian No. 2 steamer went to Fort Howard, and shortly thereafter the Fort Howard steamer responded to a fire on the north side of Green Bay.[66] Controversy did not follow. Later that summer, newspaper reports described good working relations between the fire departments, especially after a series of arson fires in Fort Howard.[67]

> *"We can see no cause for disagreement. Let brethren dwell together in unity."*
>
> - *Fort Howard Herald*

Green Bay considered buying a third steamer in 1875 after disbanding two hand pumper companies and selling those engines. The *Daily State Gazette* observed, "The city has now to rely on the two steamers in case of fire, with the assistance of [the] Fort Howard engine."[68] However, the Green Bay Fire Department Committee concluded that three steamers between the two cities were adequate.[69] Such profound confidence in assistance from across the river indicates the success of the mutual aid agreement.

Further insight can be gained from a report by GBFD Chief Scheller to the Common Council. In the second year of the mutual aid arrangement, "horse-hiring for hauling apparatus" to Fort Howard cost Green Bay $261, with Fort Howard paying $50.[70] The Green Bay aldermen accepted the report without issue. The financial arrangements were working.

A potentially controversial event occurred within two years of the agreement. At least one GBFD steamer responded to a fire alarm in Fort Howard. This turned out

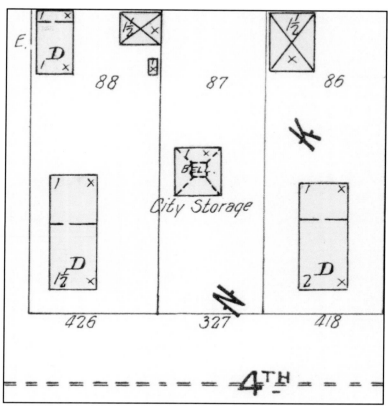

Fort Howard Resolute Fire Company shed from the 1900 Sanborn Insurance map. GBFD did not use this simple shed after the 1895 merger and is marked as "City Storage."

Mutual Aid Leads to a Merger

to be a false alarm, making the hauling expense unnecessary. This was exactly the circumstance that triggered the 1874 mutual aid crisis. In particular, nobody in Fort Howard sounded the four-tap mutual aid alarm request. Yet, the GBFD fire company crossed the bridge without official permission, in violation of department rules.[71] The *Green Bay Advocate* complained that a false alarm costs Green Bay $20-$40 and the rules must be enforced.[72]

In contrast to the disagreement two years before, the common councils and both chief engineers worked agreeably to avoid the unnecessary expenses incurred by a false alarm. They reached a new mutual aid agreement within a couple of months. Essentially, it reiterated the existing agreement from 1875.[73]

The re-established mutual aid system in 1877 worked well. Within days of the agreement, two large mutual-aid fires occurred. A sawmill adjacent to the valuable railroad depot in the south side of Fort Howard burned, with the loss of about 600,000 feet of milled lumber. Both GBFD steamers operated alongside the Fort Howard steamer from midnight until 9 am.[74] Fort Howard citizens gave all firefighters food and drink. The newspaper emphasized that "not a drop of whiskey" was provided.[75]

At another fire, a Fort Howard warehouse burned near the Main Street Bridge. GBFD Chief Scheller placed the GBFD steamer on the east side of the bridge and ordered an attack hose stretched across to Fort Howard. Although Fort Howard did not request mutual aid, the paper supported Scheller's actions, citing the need to protect the bridge.[76] At a fire at a brewery in Green Bay, the GBFD hoses froze in the 30-below-zero weather, completely disrupting operations. The Fort Howard Fire Department engine ultimately extinguished the fire.[77]

It is difficult to determine the frequency of mutual aid responses. Only two official reports survive. GBFD Chief Scheller reported to the Common Council that from April 1876 to March 1877, there had been seventeen mutual aid responses into Fort Howard—fifteen by one GBFD steamer and two requiring the entire fire department. Chief Engineer Al Gray of Fort Howard reported there were eight fires in 1883, and only one mutual aid into Green Bay.[78] A survey of the newspapers provides some insight. For most years from the mid-1850s through the mid-1880s, there were one or more mutual aid responses. There were no reports of mutual aid response at all for some years. However, the thoroughness and reliability of newspaper reporting is uncertain.

In addition to fighting fires, relations between the two fire departments prospered.

PLANING MILL AND LUMBER YARD OF A. ELDRED & SON, FT. HOWARD, WIS.

Eldred planing mill and lumber yard from an 1881 illustrated atlas. A massive fire destroyed all of the buildings and much of the piled lumber in September 1883. The entire GBFD responded to assist the Fort Howard Fire Department, but the fire overwhelmed their combined efforts.

In addition to fighting fires, relations between the two fire departments prospered. Newspapers reported that Fort Howard Fire Department participated in the July 4, 1876, parade in Green Bay as well as the 1877 and 1886 GBFD annual inspection days.[79] The historic Final Review of the volunteer GBFD in September 1891 included Fort Howard. In fact, the Fort Howard hose company won the "first to water" competition that day.[80] Members of GBFD also attended the second annual Fort Howard

Fire Department dance on December 28, 1876.[81] The firefighters from both cities almost certainly attended each other's functions much more often than was reported by the newspapers.

The positive relationship between the fire departments resulted in other benefits. The Fort Howard steamer responded to all alarms in Green Bay while GBFD Enterprise No. 1 steamer was out of service for maintenance during a few days in May 1879.[82] A few months later, GBFD reciprocated when the Fort Howard steamer was out of service two weeks for repair.[83] During this time, GBFD "promptly" responded to a machine shop fire in Fort Howard. Green Bay firefighters successfully protected adjacent structures, although the first building was lost.[84] Fort Howard repeated the favor in August 1883 when Guardian No. 2 steamer underwent a month-long boiler overhaul.[85]

The historical records reveal a curious connection between the two fire departments. Beginning in 1874, the Green Bay Common Council frequently paid John Kittner $4-$10, most of these payments likely were for "fireing the steamer" (igniting the fire in the steamer burn box).[86] He was a member of GBFD Germania No. 1 and the "stoker" for Enterprise No. 1 steamer.[87] The last payment by Green

Bay to Kittner was in August 1877.[88] The very next month, Fort Howard paid Kittner for "fireing the engine" for its fire department.[89] Fort Howard made even more payments to Kittner until March 1880.[90] This strongly suggests he had become a member of the Fort Howard Fire Department.

However, Kittner returned to GBFD by September 1880, serving as the assistant foreman for Germania No. 1.[91] Kittner remained with GBFD and became one of the first full-time, paid firefighters with the transition from volunteer to full-time in 1890-92.[92] He resigned from GBFD in June 1892.[93] Thus, having served first with GBFD, then Fort Howard Fire Department, then again Green Bay, Kittner became the first area firefighter documented to "swing" across the river and was the first genuine Green Bay firkle.*

An incident at a fire in 1883 reveals just how well the cities and fire departments cooperated. A large fire occurred at the Eldred Lumber and Manufacturing Compa-

---

To "swing" is modern Green Bay Metro Fire Department parlance for a firefighter to move from one station to another. A "firkle" is a firefighter who shows extreme, sometimes excessive, enthusiasm for the job. It is derived from the term "fire call." Upon hearing the fire alarm, the overly spirited firefighter will speak rapidly, and the words "fire call" become blended: fire call . . . fir-call . . . firkle. It is simultaneously an insult, a compliment, and a point of pride.

ny in Fort Howard, located along the Fox River at what is now Broadway and Tenth Street.[94] Friction from a pulley threw a spark into a pile of shavings, which caught fire. The blaze quickly spread, destroying the planing mill, sheds, massive piles of shingles and lumber, five railroad cars, and even logs in the river. About six million feet of lumber were ruined and the total loss was approximately $175,000.[95]

The entire GBFD responded. At one point, a horse-drawn GBFD hose cart attempted to move further into the burning lumber yard. The cart struck a pole, caught a wheel, and the horses fell, becoming ensnared. Firefighters cut the horses free of the harnesses. The two horses and two human rescuers were burned, though all survived. The flames also damaged the GBFD hose cart and hose.[96]

Officials initially estimated damages to be $218.50 for GBFD equipment and $375 for the two horses owned

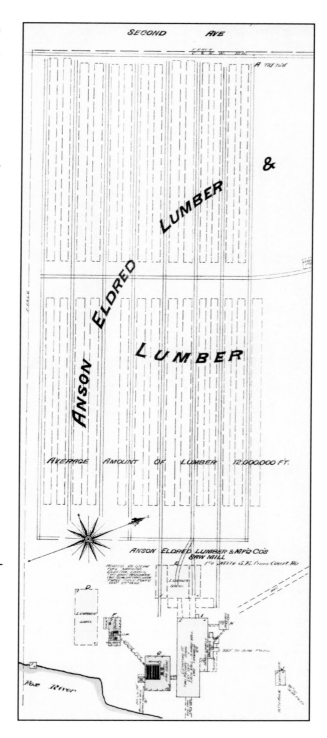

Anson Eldred Lumber and Manufacturing Company from the 1883 Sanborn Insurance map. Fire destroyed much of the facility in 1883, located between the Fox River (bottom) and modern South Broadway (top) at modern Tenth Street. A note indicates an average of twelve million feet of lumber on site. Half this amount was destroyed in the fire. A GBFD hose cart became trapped, resulting in damaged equipment and burned horses. Compensation from the City of Fort Howard came quickly, avoiding controversy and highlighting the good mutual aid relationship between the fire departments.

Mutual Aid Leads to a Merger

by Frank Hagen of Green Bay.⁹⁷ Within three months (fast for municipal government), Fort Howard paid $200 for the equipment and $150 to Hagen.⁹⁸ The Green Bay Common Council thanked the Fort Howard aldermen for resolving this issue.⁹⁹ Furthermore, the Green Bay aldermen directly paid Hagen $125 and returned $25 to Fort Howard.¹⁰⁰ Unlike 1874, the cities and departments readily resolved this financial issue without difficulty.

The joint municipal water works also brought the communities together. Fear of a repetition of the Great Fire of 1880 motivated Green Bay municipal, civic, and business leaders to establish a municipal water works. The system provided the dual benefits of safe, drinkable water and abundant supply for firefighting on both sides of the river. The two communities readily approved formation of the joint Green Bay and Fort Howard Water Works Company.¹⁰¹ Use of each city's name reveals the movement toward

Political cartoon from the April 4, 1895, *Green Bay Advocate*. This image portrays Green Bay as the groom and Fort Howard as the bride reaching across the Fox River.

a single community. The system became fully functional and officially accepted in both communities by July 1887.

The separate and parallel existence of Fort Howard and Green Bay eventually brought about a predictable outcome. In April 1895, voters overwhelmingly approved annexation of Fort Howard by Green Bay: 1,631 for and 60 against in Green Bay, 930 for and 154 against in Fort Howard.[102] A massive celebration erupted on both sides of the Fox River with bonfires, fireworks, cannons booming, and people dancing and marching in the streets.[103] City administrators established April 2, 1895, as the official date that Fort Howard no longer existed and "Greater Green Bay" was created.[104]

The annexation ordinances spelled out specific changes to fire protection on the west side.[105] First, the pole-mounted, sidewalk-level fire alarm system would be extended to the west side. The ordinances required at least fifteen fire alarm boxes. Fort Howard had considered, and rejected, an alarm system shortly after Green Bay installed its system in 1892.[106] Next, two new fire stations would be erected, one on Broadway near downtown and the other in the most southern west-side ward. The fire department would equip each with hoses, carts, and other equipment.

IT IS DONE

The Union an Accomplished Fact

GREATER GREEN BAY

And a Night Of Enthusiasm.

Headline celebrating the approval of the annexation of Fort Howard by Green Bay from the April 4, 1895, *Green Bay Advocate*. After several attempts over four decades, voters on both sides of the river overwhelmingly approved the measure. An epic party ensued in both cities. Though this combined the municipal governments, in reality the two fire departments had been functioning together very well for almost forty years.

Mutual Aid Leads to a Merger

Changes to the fire department on the west side occurred quickly. The Common Council authorized the Fire Department Committee to "make necessary improvements in No. 3 Engine House, West Side, and to equip the same properly."[107] This was the first official reference to the west-side station as GBFD Station No. 3. The very next day (May 4, 1895), GBFD Chief William Kennedy took over the engine house on South Pearl Street.[108]

A newspaper report described the Fire Department Committee as "pushing the improvements in the fire department on the West Side as rapidly as possible."[109] They cleaned out the upstairs room formerly used by Fort Howard municipal officials and prepared a sleeping area, including four cots. Other station improvements included a hole cut in the dorm floor for a sliding pole to the apparatus bay, coating the walls for a better appearance, improved horse stalls, and widening the apparatus bay door to accommodate a larger hose cart. Also, a new telephone line was installed to improve communication with the Water Works. The old mechanism frequently failed in bad weather.[110]

The West Side now has Sure-Enough Firemen.

The engine house on the west side was Saturday taken possession of by Chief Kennedy and became a regular station with real, sure-enough paid firemen. During the day a sleeping apartment was arranged and cots placed in position for four men, and from now on they will be at their posts day and night ready to answer the summons that is almost always a call to danger and hard work. The new firemen engaged are all west side men, and are as follows: Theodore Leicht, foreman; John Walters, teamster; and Martin Cleary and Chas. Rasmussen.

Detailed announcement on the newly occupied GBFD west side engine house from the May 9, 1895, *Green Bay Advocate*. The four men hired are listed as west siders and two (Leicht and Walters) can be confirmed as former Fort Howard Fire Department volunteers.

Most significantly, the Green Bay Fire Department Committee appointed four men as full-time, paid firefighters. They began service the same day the engine house was taken over.[111] The *Green Bay Advocate* proudly announced the former Fort Howard engine house "became a regular fire station with real, sure-enough hired firemen," and they "will be at their posts day and night."[112] For the first time, the west side had full-time firefighting coverage staffing GBFD Hose Company No. 3.[113] Fort Howard firefighters previously were volunteers from 1856, then paid, part-timers beginning in 1883.

All four of the new GBFD personnel were "west side men," and at least two (Theodore Leicht, captain, and John Walters, driver) can be confirmed as former Fort Howard firefighters.[114] Leicht had served as treasurer.[115] Walters had worked as the Fort Howard city teamster since 1893. He used the city horse team to haul the steamer when not otherwise doing street work.[116] The Fire Department Committee's June 1895 report showed an increase salary cost, reflecting four new salaries.[117] Their average salaries were $40 per month. The 1896-97 *City Directory* for the unified Green Bay lists GBFD as having sixteen members in three hose and one hook-and-ladder companies.[118] Four of these firefighters and Hose Co. No. 3 were on the west side.

THE WEST SIDE FIRE COMPANY.

FOUR SALARIED MEN NOW ON DUTY OVER THERE.

Theodore Leicht Receives the Appointment as Captain—More Hose to be Purchased and Changes Made in the Engine House—Firemen Will Sleep in the Old Council Rooms—Electric Fire Alarm Will Soon be Put In.

Assignment of full-time firefighters announced in the May 4, 1895, *Green Bay Gazette*. GBFD took over the former Fort Howard volunteer fire station on South Pearl on May 4, 1895. The department immediately made renovations, and most importantly, four full-time firefighters went on duty.

New fire alarm box locations on the west side from the August 29, 1895, *Green Bay Advocate*. The annexation ordinance required installation of alarm boxes in the former Fort Howard. Pulling the alarm box handle sent a telegraph signal to gong bells at all three GBFD stations. The box number listed matched the sequence of the gongs. For instance, Box No. 124 at Broadway and School Place would trigger a sequence of one gong, followed by a pause, two gongs, another pause, and then finally four gongs. All the new west side alarm boxes had three digits, starting with "1" while the east side alarm boxes were only two digits.

FIRE BOX LOCATIONS.
As Established and Announced by Chief Kennedy.

The following locations for alarm boxes on the west side have been fixed by Fire Chief Kennedy.

BOX NO.	LOCATION.
112	Corner Broadway and Dousman streets.
113	Corner Dousman and George streets.
114	Corner Broadway and Elmore streets.
115	Corner Cedar and Mather streets.
116	McDonald's mill.
121	Corner Main and Willow streets.
123	Engine house No. 3.
124	Corner Broadway and School place
125	Corner Second street and Third avenue.
126	Corner Third street and Fifth avenue.
131	Corner Second street and Eighth avenue.
132	Corner Fifth street and Seventh avenue.
134	Corner Fifth street and Third avenue.
141	Corner Eighth street and Third avenue.
142	Fort Howard Lumber Co.—Office.

Other essential changes followed. GBFD Truck No. 1 and Hose Cart No. 1, housed at the South Washington station, now automatically went to any fire on the west side.[119] Chief Kennedy requested a light carriage so he could quickly respond to west-side calls.[120] By the end of May, Green Bay purchased 1,500 feet of new cotton hose for Station No. 3.[121]

The final change to the fire department occurred at the end of August 1895, about four months after annexation. Chief Kennedy presented a list of locations for the fifteen new alarm boxes on the west side.[122] The city selected the Gamewell Company system, the same manufacturer as the east side system that had been installed in 1892.[123] The expanded alarm system cost $3,000 and was interconnected to the east side as well as the Water Works facility.[124]

Though not a direct alteration to the fire department, the Water Works became the Green Bay Water Company less than a year after annexation.[125]

As a sort of finale for the Fort Howard Fire Department, the Green Bay Common Council authorized payment of $202.72 to "the fire laddies of the department for [the] West Side."[126] This was the last payout to the part-time Fort Howard firefighters. That fire department was now gone.

Merging the fire departments notably changed fire protection in the former Fort Howard. However, the city did not entirely fulfill the annexation ordinances. The agreements clearly required construction of two fire stations, one on Broadway and one in the former South Ward.[127] However, GBFD simply took over the former Fort Howard station on South Pearl and stayed there. This station remained in service until 1937, when new Station No. 3 opened on Shawano

FIRE ALARM BOXES.

EAST SIDE.
- 12—Cor. Main and Washington.
- 13—Cor. Main and Jefferson.
- 14—Engine House No. 1, Washington St.
- 15—Cor. Doty and Jefferson.
- 16—Cor. Adams and Cherry.
- 21—Cor. Adams and Crooks.
- 23—Cor. Jefferson and Mason.
- 24—Cor. Monroe and Cass.
- 25—Cor. Monroe and Eliza.
- 26—Cor. Jackson and Mason.
- 31—Cor. Crooks and Monroe av.
- 32—Cor. Crooks and Webster av.
- 34—Cor. Monroe av. and Cherry.
- 35—Cor. Monroe av. and Cedar.
- 36—Cor. Main and Jackson.
- 41—Cor. Van Buren and Walnut.
- 42—Engine House No. 2, Main St.
- 43—Cor. Walnut and Eleventh.
- 45—Cor. Doty and Twelfth.

NORTH SIDE.
- 51—Cor. Harvey and Jackson.
- 52—Cor. St. George and Main.
- 53—Cor. Main and Pleasant.
- 54—Cor. Harvey and Twelfth.
- 61—Murphy Lumber Co. planing mill.
- 62—Murphy Lumber Co. office.
- 63—Hagemeister Brewery.

WEST SIDE.
- 112—Cor. Broadway and Dousman.
- 113—Cor. Dousman and Norwood av.
- 114—Cor. Broadway and Elmore.
- 115—Cor. Ashland av. and Mather.
- 116—McDonald's Mill.
- 121—Cor. Walnut and Oakland av.
- 123—Engine House No. 3.
- 124—Cor. Broadway and School.

SOUTH SIDE.
- 125—Cor. W. Mason and Broadway.
- 126—Cor. Broadway and Fifth.
- 131—Cor. Mason and Greenwood av.
- 132—Cor. Fifth and Oakland av.
- 134—Cor. Maple av. and Third.
- 141—Cor. Eighth and Broadway.
- 142—Fort Howard Lumber Co. office.

HOW TO SOUND ALARM OF FIRE.

Notice:—Keys to boxes can be found at the four houses nearest the box and are carried by all policemen. In case of fire, go to your nearest box, get key, unlock box, pull hook on inside *Once* only, and let go. *Stand By the Box* until the department arrives and direct them to the fire.

To send alarm by telephone, call up Central Telephone Exchange, who will repeat to headquarters.

SPECIAL SIGNALS FOR COMPANIES.

Hose Cart No. 1, 2—1 and box. Hose Cart No. 2, 2—2 and box. Hose Cart No. 3, 3—3 and box.

3 taps of bell after alarm, more pressure.
General alarm, 9—1 and box.
Fire out, 5 taps.
Fire alarm test, 1—1.

To locate a fire, count the blows on the bell as they are struck; thus, for box 13, 1—3; for box 25, 2—5; for box 125, 1—2—5.

Horses must all stand hitched 30 minutes after the first alarm.

List of all fire alarm boxes in Greater Green Bay from the 1896-97 *City Directory*. The instructions state that the boxes had to be opened with keys kept in the four nearest houses or carried by policemen. Other special signals state how the firefighters can communicate back to the fire stations or the Water Works Company.

Mutual Aid Leads to a Merger

Avenue at Hazel Street, forty-two years after annexation. Even later, Station 4 opened on Ninth Street at Chestnut in 1949, finally fulfilling the 1895 requirement for a second west-side station. It is unknown why these obligations took so long to fulfill.

In contrast to other steps in the development of GBFD, major fires did not lead to the union with Fort Howard. Rather, this merger resulted from an almost evolutionary process over four decades.

Cooperation between the Green Bay and Fort Howard fire departments began as simple mutual aid. As responses became more frequent, officials established formal agreements, ensuring continued support and eventually culminating in a complete merger. For the most part, social, commercial, and governmental motives induced annexation. However, to an immeasurable extent, cooperation between the fire departments was also a factor. Thus, the success of mutual aid was a partial force of change that placed the Green Bay Fire Department on both sides of the Fox River.

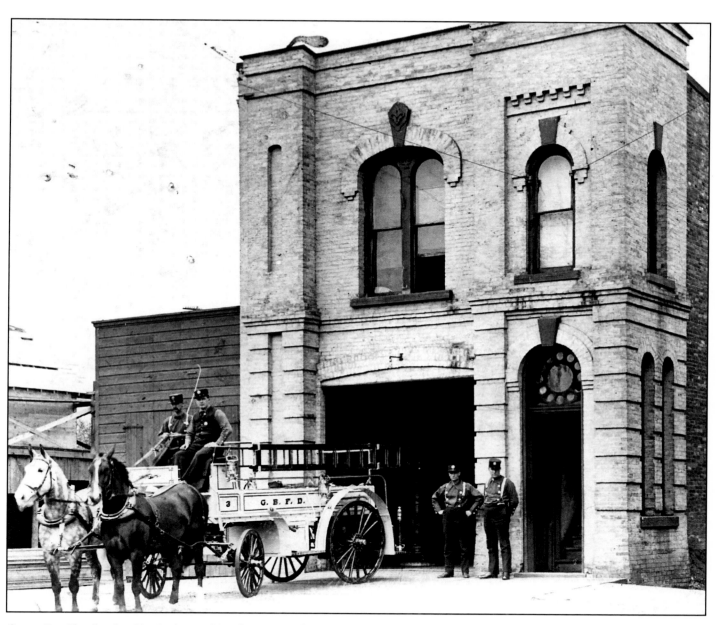

Green Bay Fire Station No. 3 about 1900, five years after the merger. The City of Fort Howard built this station in 1873 for the volunteer fire department to replace one destroyed by fire. Originally it featured a bell tower above the open doorway. The hose cart is marked with the company number "3" and "G.B.F.D." This house served as GBFD Station 3 until 1937, when it was replaced by current Station 3 on Shawano Avenue.

Recreation of the Franklin No. 3 Hose Co. house image from 1890 (page 160) with modern Green Bay firefighters in August 2014. The Franklin No. 3 house on Main Street at North Irwin Avenue stopped serving as a fire station in 1892. It had various functions there until it was moved to Heritage Hill State Historical Park in 1975

Epilogue

Ideally, a single volume would cover the entire history of the Green Bay Fire Department (GBFD). But most things, including books, must have limits. This book describes the events from 1836 to 1895, and extending the narrative another 120 years to the present would have been physically impractical.

The merger with Fort Howard Fire Department, with the city and its full-time fire department now on both sides of the Fox River, is an appropriate point to end this volume. The early years of volunteer fire companies symbolize GBFD's infancy as an organization. GBFD then entered a sort of adolescence period as a smaller and less-developed version of today's department.

In 1895, GBFD consisted of sixteen firefighters in three stations, with three hose companies and one hook-and-ladder company. As of 2016, the Green Bay Metro Fire Department consists of nearly 200 firefighters in eight stations, with seven engine, two ladder, two battalion chief, and five paramedic ambulance companies. Growth of the department matched growth of the city, but size was not the only change. There have been dramatic alterations in vehicles, equipment, tactics, and services—all of which came about because of equally fascinating forces of change.

I've been asked frequently why I embarked on this project. Very simply, it was because I made a series of erroneous assumptions. In 2013, department members proposed creating an updated edition of the 1991 GBFD Legacy Album, a sort of fire department yearbook. When subjects were offered, I took the department history topic thinking, "How hard can it be?" The history chapter in the 1991 edition mostly drew from a series of *Green Bay Press-Gazette* articles written in the 1960s by Jack Rudolph, the newspaper's local history columnist. That chapter was a straightforward, chronological narrative of the fire department history. It provided glimpses into the past, and I figured all I needed to do was conduct some simple research to fill in the back stories. I was wrong.

The pivotal moment occurred while conducting research at the Area Research Center on the Seventh Floor of the Cofrin Library at the University of Wisconsin-Green Bay. The center serves as a repository for area municipal records, including Green Bay Common Council minutes dating to 1853. From the 1991 Legacy Album history chapter, I knew that 1891 was an important year, so I explored those records. In a quirk of fate, I opened the 1891 Common Council minutes to page 490, exactly where the ordinance was recorded establishing the drastic reorganization of GBFD. Three hand-written pages describe the complete revamping of the fire department. The next question was obvious: What brought this about?

After pouring through the Common Council minutes, newspaper microfilm, and existing volunteer fire company records, the story of the disastrous 1890 Thanksgiving Britton fire and subsequent fallout became clear. This was a fascinating story. Previously, I had presumed the conversion to a full-time department came about simply because it seemed like a good idea at the time. This, too, proved wrong. It turned out the change to full-time status included a captivating story that answered the fundamental historical question, "Why did this happen?"

As the research continued, I found that every major change to GBFD had an intriguing explanatory story. Nothing just happened. Every significant change was a consequence of major fires. I found this profoundly fascinating.

To quote the fictional Captain Jean-Luc Picard of *Star Trek*, "A mystery is irresistible. It must be solved." That's what happened to me. That's why I took on this project.

This book is about changes to GBFD. More specifically, it presents a clean story highlighting successful, positive transformations. But certainly not everything went as planned. There had to have been many other changes suggested and implemented that ultimately failed and were rejected. These typically were not well documented, and thus have been lost to history.

The theme of this book has been change. It actually could be the theme of any history of the fire service. Everything we do is different than before. I have witnessed significant alterations to operations just since my career began in 1997. We firefighters don't like change, but it is inevitable. I don't suggest we blindly embrace change. Our scrutiny is needed. We must continually sift and winnow through the good and bad ideas routinely offered to the fire service. Most importantly, we should never accept the status quo by saying, "That's the way we've always done it."

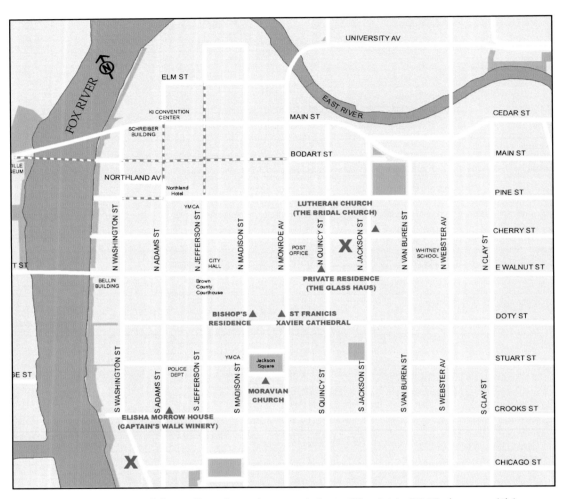

1880 survivor map. Buildings directly endangered that still exist in 2016 shown within modern Green Bay. Triangles mark the locations of these buildings. The X at South Washington and Chicago shows the fire origin at the Astor Planing Mill, while the X on Cherry between Quincy and Jackson marks the origin of the northern section of the conflagration. Streets that no longer exist as of 2016 are indicated by dashed lines.

Appendix I

Survivors of the Great Fire of 1880

The Great Fire of September 20, 1880 (described in Chapter 6), destroyed about one hundred buildings in Green Bay, including sixty homes and one church, with the remainder being smaller residential outbuildings. Many more buildings were directly threatened, but survived through the efforts of firefighters and citizens.

Six of these structures still stand as of 2016 and are pictured on the following pages. All six were directly in the path of the conflagration and were inundated with burning debris carried by the strong wind.

Next-day headline from the September 21, 1880, *Daily State Gazette*, a Green Bay newspaper.

1880 Survivors   287

Elisha Morrow house at 324 South Adams (above). From the cupola of this house (built in 1857), Morrow dumped buckets of water to extinguish burning debris landing on the roof. He watched as the conflagration progressed around his home, which is currently the Captain's Walk Winery.

Moravian Church at 518 Moravian (right). This church was built in 1851 on the south side of Jackson Square. GBFD firefighters successfully stopped the southern section of the conflagration on the west side of the square, about one-half block away. The church was moved to Heritage Hill State Historical Park in Allouez in 1980.

Bishop's Residence at 139 South Madison (above). This was built in 1870 and is located one block north of Jackson Square, where firefighters stopped the southern section of the conflagration. GBFD Enterprise No. 1 steamer drew water from an underground cistern in front of the Bishop's Residence.

St. Francis Xavier Cathedral at 140 South Monroe (left). This church is located one block north of Jackson Square and was completed shortly before the fire.

1880 Survivors

Private residence at 633 East Walnut. This private residence (built about 1871) is on the block adjacent to the Charles Kitchen home, where the northern section of the conflagration began. Today it is the Glass Haus shop.

Lutheran Church at 901 Cherry. This church (built in 1863) is less than two blocks from the Kitchen home, where the northern section of the conflagration began. Fifteen homes burned in this section of Cherry, but the church was saved. Currently, it is the Bridal Church.

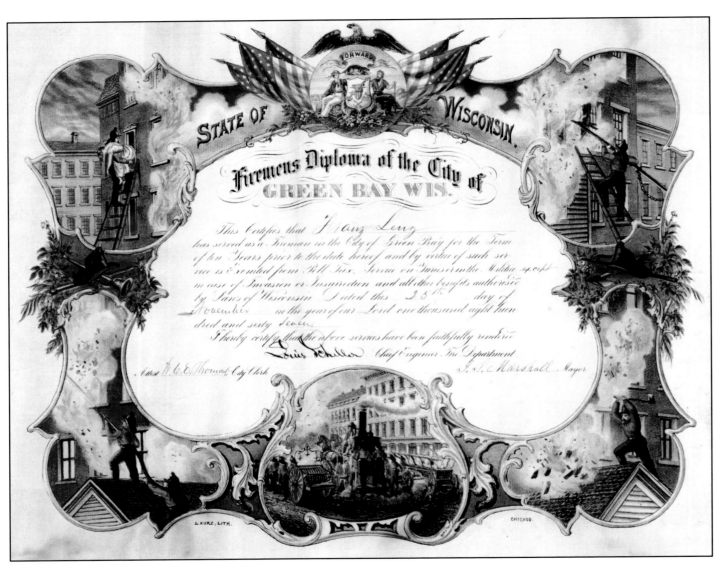

Franz Lenz's GBFD Firemen's Diploma from 1867 in the private collection of Rick Fleury. The certificate acknowledges exemption from poll tax, jury duty, as well as militia service and was given after ten years of fire department service.

Appendix II

Artifacts of the Early Green Bay Fire Department

Old Croc hand pumper. The US military provided this fire engine (built by Harry Ludlum of New York in the 1820s) to the Fort Howard garrison. It was then sold to the Borough of Green Bay in 1843 and used by several GBFD companies until it became obsolete. Local businesses later used Old Croc for a while, but then it was stored behind a fire station. Veteran volunteer firefighters paraded with Old Croc as part of the last volunteer review in September 1891, captured in the image on page 187. Eventually, it became a part of the Neville Public Museum collection, as shown here in the 1930s or 1940s.

All items are in the Neville Public Museum collection except where noted.

James Smith manufactured hand pumper with Algoma Fire & Rescue Association. The Borough of Green Bay purchased this fire engine in 1851. Several GBFD companies used it until it was sold to the Ahnapee (now Algoma, Wisconsin) Fire Department in 1875. It was later sold to the Luxemburg Fire Department, but eventually returned to Algoma.

Hose cart with Algoma Fire & Rescue Association (right). Green Bay sold a hose cart to Ahnapee (Algoma) in 1875 along with the Smith hand pumper. It is believed this is the same cart, though there is no direct evidence.

Hose carts at Heritage Hill. These two hand-drawn hose carts were in GBFD storage until donation to Heritage Hill in the 1990s. The history of these carts is unknown. Firefighters pulled these carts by hand and steered with the bars extending from the frames. They deployed fire attack hose from the large reel between the wheels.

Franklin Hose Company No. 3 house at Heritage Hill. The city built this in 1887 on the south side of Main Street at the intersection with modern North Irwin. It closed in 1892 when the fire department transitioned from volunteer to full-time. The building later served as a voting place, band practice room, meeting hall, and public library until it was moved to Heritage Hill State Historical Park in 1975. It is pictured here during a reunion with current Green Bay firefighters in 2014.

Artifacts

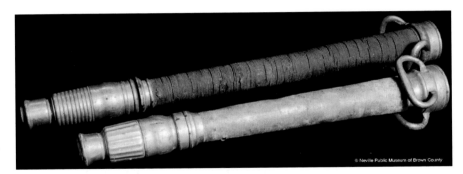

Fire nozzles (right). These 2½-inch, smooth-bore nozzles feature internal gate-valves in the forward sections that open and close by twisting the housing.

Guardian No. 2 banner. A group of Green Bay ladies gave this banner to the fledgling Guardian No. 2 fire company as part of the July 4, 1858 celebrations. It can be seen in the 1870 image on page 42.

Speaking trumpets. These two trumpets are identical in material and detail to the trumpet given to Chief Engineer Lindley by the steamer manufacturing company (page 298, lower right). Company foremen and other officers likely used these.

Lamp from Guardian No. 2 steamer (above). The faces of this lamp feature the company name, "wide awake" motto, and an open eye. This lamp can be seen on the steamer in the 1876 image on page 79.

Speaking trumpet (above). G. W. Hannis Company of Chicago gave this trumpet as part of the sale of the new apparatus to Washington Hook & Ladder Company No. 1 in 1876. It is inscribed (left) with "Foreman, W. H. & L. Co."

Artifacts 297

Alarm boxes (right). These recently restored boxes are from the family of GBFD Chief David Zuidmulder. The two boxes to the left are from the 1920s, the white handle alarm box from the 1950s, and the terminal box to the right from the 1920s.

Prize trumpet. Guardian No. 2 won this trumpet (also on page 46) for generating the longest hose stream of the three Green Bay and one Fort Howard hand pumper companies during the 1860 Brown County Firemen's Tournament. The detailed work is in great condition, suggesting this was not used at fire scenes.

Speaking trumpet. The Clapp and Jones Company gave this trumpet to the fire department as part of the sale of Guardian No. 2 steamer. It is inscribed with "Samuel Lindley, Chief Engineer, 1872." The chief engineer was in overall command of the fire department, equivalent to the modern term fire chief. Those in charge used trumpets to facilitate communication of commands at fire scenes.

Engine House No. 1 bell in Amberg, Wisconsin. The city purchased this 3,000 pound bell in 1874 for the South Washington engine house. It was removed in 1925 and donated to St. Joseph Church in Green Bay. Eventually, it was given to the St. Agnes Church in Amberg, where it is rung at noon and to announce services. Embossed on the side is "The Jones & Company Troy Bell Foundry, Troy, N. Y., 1874," which matches Green Bay Common Council records.

Speaking trumpet at Heritage Hill. This trumpet is similar in material and design to the trumpets given to the fire department in 1872 as part of the Guardian No. 2 steamer sale (page 298, lower right). This history of this piece is unknown.

Artifacts

Keystone from Engine House No. 1 in the lobby of the Backstage at the Meyer. In 2014, renovators digging at the site of the former engine house on South Washington found this piece buried, likely used as back-fill of an old basement. It was under concrete flooring dated to about 1929, the year the fire station was demolished. In the top image from about 1925, two keystones are seen in the upper, arched windows (closeup of highlighted section). The shape of the keystones used in Station No. 1 match perfectly with the one discovered. Keystones provide structural support at the top of a brick arch.

Gamewell Fire Alarm system ticker tape machine. Signals from the alarm boxes activated the alarm gong bell, but also triggered this machine to print the alarm box number, which indicated the street location.

Washington Hook and Ladder No. 1 belt. The "Hooks" used this belt to carry tools. Modern firefighters use similar, more sophisticated, ladder belts.

Gamewell Fire Alarm system alarm box on stand. This restored box was made between 1900 and 1916. The number 28 on the box front below the word "Station" (closeup on left) indicates the location, which in this case was East Mason and Roosevelt. Opening the door revealed a handle, which triggered a telegraph signal when pulled.

Artifacts

1892 Gamewell Fire Alarm system control board. This electronic mechanism, known as a repeater, controlled the telegraph signals between street-level alarm boxes and the gong bells at the fire station. The glass case protected the exposed wiring from dust and water. This piece was part of the first system installed in Green Bay in 1892.

Gamewell Fire Alarm gong bell (below). The signal from the alarm box would trigger this gong bell, also known as a regulator, to sound in a sequence matching the specific alarm box number. Box 18 would give a single gong, a pause, then eight gongs. The sequence was repeated a few times. Firefighters then knew to respond to Box 18, which was at the corner of Adams and Pine.

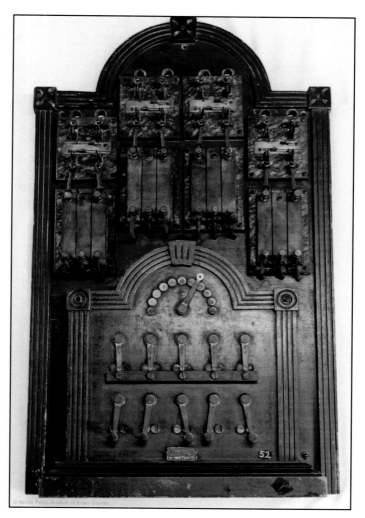

Gamewell Fire Alarm system control board (above). The levers on this board were used to regulate the electronic connections of the system. In addition to sounding the gong bell, the same signals automatically opened the doors of the horse stalls on the apparatus floor.

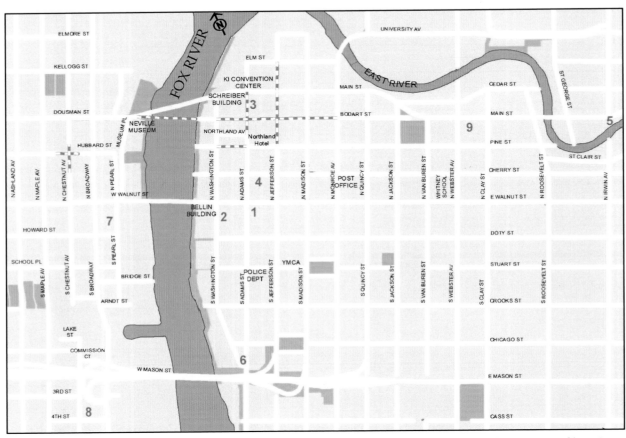

Former fire station locations within modern Green Bay. Numbers on the map correspond to images of location within this appendix. Streets that no longer exist as of 2016 are indicated by dashed lines.

Appendix III

Current Views of Former Fire Station Sites

1

Engine house at the southeast corner of Adams and East Walnut. The borough built the first engine house in Green Bay here in about 1842. It was used until about 1852.

2

Fire stations at 111-115 South Washington just south of East Walnut. There have been two fire stations on this site, which was within the three window section of the Backstage at the Meyer, immediately adjacent to the theatre. The borough built a wooden engine house here in about 1852, which was used by Alert and Germania No. 1 companies. In 1868, the wooden building was moved and a single apparatus bay, brick-exterior station built for the new Enterprise No. 1 steamer. The city added a second apparatus bay in 1883 for Washington No. 1 ladder company. It was later designated Station 1 and served until 1929.

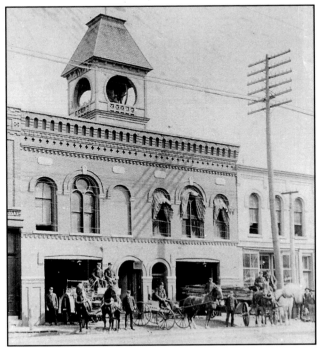

3

Guardian No. 2 house at 408 North Adams. This site is within the current Schreiber Foods headquarters complex. North Adams and former Main Street met at the corner of modern building, which corresponds to corner of the brick building from the 1881 image. In both images, former Main Street goes from left to right. The engine house (visible left center in the 1881 image) was about 140 feet beyond the building corner, behind the dark wall, in between the Schreiber building and the parking garage. The engine house was built in 1859, renovated in 1883, and remained in service until 1892.

Current Views of Former Fire Station Sites

4

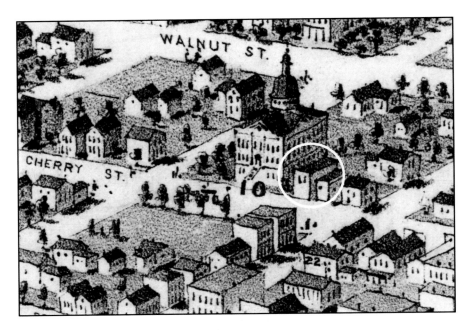

Washington Hook and Ladder No. 1 house at 314 Cherry. The "Hooks" house was built in 1860, and used by this company until they moved into the addition to the station on South Washington in 1883. The rather narrow ladder house fit within the parking lot shown here, in the center between the pale brick wall and the driveway.

Franklin No. 3 house on the south side of Main Street at North Irwin. Two Franklin No. 3 houses were located at this site. The first was built in 1860 and then shut down when the company was disbanded in 1875. The second house was built in in 1887 and served until 1892. The fire company houses were located in the driveway shown here.

5

Current Views of Former Fire Station Sites

6

Astor No. 1 house at the southern end of the triangular park at Washington and Adams. This engine house was originally on the South Washington site near Walnut. The city moved it to this location in 1868 to make room for a new, larger station. Astor No. 1 responded from this house using the Smith hand pumper until it was disbanded in 1875. Modern Green Bay Metro Fire Department Station No. 1 (opened 1929) occupies the northern section of this block. The Astor house was located on the lawn adjacent to the concrete apron.

7

Fort Howard volunteer station and GBFD Station No. 3 at 118-120 South Pearl. The Borough of Fort Howard built a station for their volunteer fire department in 1858 and moved it to this site in 1860. Fire destroyed that building in 1872 and the brick exterior station was built as a replacement. This station served the volunteer Fort Howard Fire Department until annexation by Green Bay in 1895. It then became GBFD Station 3 until it was closed in 1937. As of 2016, this is an abandoned commercial building.

Current Views of Former Fire Station Sites

8

Resolute Fire Company site at 418 Fourth Street. Fort Howard authorities placed a fire alarm bell in this southern ward location in 1872. They later located a hose cart in a simple shed in 1887. One year later, Resolute Fire Company, part of the Fort Howard Fire Department, began to respond with this hose cart. It remained in service until annexation by Green Bay in 1895.

9

GBFD Station 2 at 1018 Main Street. The city built this station in 1892 as part of the sweeping modernization efforts. It served until 1965.

Current Views of Former Fire Station Sites

Green Bay Firefighter Tribute on the west bank of the Fox River, near the Neville Public Museum (right). Surrounding the monument (dedicated in 2008) are bricks memorializing the service of all known Green Bay firefighters installed upon the end of their service. Firefighters wear the emblem at left during periods of mourning.

Appendix IV

Green Bay Fire Department Line-of-Duty Deaths

There is no strict definition of a firefighter line-of-duty death. Originally, it included only those who died from trauma suffered while actively fighting a fire. Other causes have been recognized more recently. The definition has evolved to include cancer and heart disease that developed while serving as a firefighter. The reasoning is that firefighters are exposed to a range of toxic chemicals in smoke despite following safe practices and use of good equipment.

For purposes here, line-of-duty death applies to any firefighter who died at any time while on duty, while involved in a fire department response, or cancer while still with GBFD.

Hans Hansen. Died February 10, 1892, four days after being injured when the horse-drawn hose cart he was driving overturned while responding to a fire.

Fred Mathews. Died January 22, 1939, probably of a stroke while returning to Station 1 in a ladder truck.

August Joppe. Died May 3, 1940, of a cardiac condition while sleeping at Station 1.

Frank Nooyen. Died January 5, 1943, of a heart attack while working at a fire in the 100 block of North Pearl Street.

Arthur Christensen, Sr. Died April 30, 1945, of sarcoma.

Fred Rein. Died September 8, 1960, of injuries when his personal car slipped off a jack stand behind Station 3.

Dan Kennedy. Died June 12, 2003, of cancer.

Arnie Wolff. Died August 13, 2006, of smoke inhalation after floor collapse at a residential basement fire on Edgewood Drive.

Mike Miller. Died June 20, 2015, of cardiac arrhythmia at night at Station 3.

Bibliography

ABBREVIATIONS

BCL: Brown County Library, Local History and Genealogy Department, Green Bay, Wisconsin

NPM: Neville Public Museum of Brown County, Green Bay, Wisconsin

UWGB-ARC: University of Wisconsin Green Bay-Area Research Center, Cofrin Library, Green Bay, Wisconsin

BOOKS

Adams, Diane L., John F. Graf, and Dell G. Rucker. *On Washington Street: A Photographic Memory*. Green Bay: NPM, 1994.

Beers, J. H. *Commemorative Biographical Record of the West Shore of Green Bay, Wisconsin: Including the Counties of Brown, Oconto, Marinette and Florence, Containing Biographical Sketches of Prominent and Representative Citizens and of Many of the Early Settled Families; Illustrated*. Chicago: J. H. Beers, 1896.

Bella French ed. *The American Sketch Book: History of Brown County, Wisconsin*. Vol. 3. Green Bay: American Sketch Book, 1876.

Conway, W. Fred. *Chemical Fire Engines*. New Albany, IN: Fire Buff House, 1987.

Conway, W. Fred. *Those Magnificent Old Steam Fire Engines*. New Albany, IN: Fire Bluff House, 1997.

Dunshee, Kenneth Holcomb. *Enjine!-Enjine! A Story of Fire Protection*. New York: Harold Vincent Smith for Home Insurance, 1939.

The Encyclopedia of American Hand Fire Engines. Handtub Junction, USA, 2001.

Foley, Betsy. *Green Bay: Gateway to the Great Waterway*. Woodland Hills, CA: Windsor Publications, 1983.

History of Northern Wisconsin: An Account of Its Settlement, Growth, Development and Resources; An Extensive Sketch of Its Counties, Cities, Towns and Villages. Vol. 1. Chicago: Western Historical, 1881.

Illustrated Historical Atlas of Wisconsin. Chicago: H. R. Page, 1881.

Martin, Deborah B. *History of Brown County Wisconsin: Past and Present*. Chicago: S. J. Clarke, 1913.

Rudolph, Jack. *Birthplace of a Commonwealth: A Short History of Brown County, Wisconsin*. Green Bay: Brown County Historical Society, 1976.

Rudolph, Jack. *Green Bay: A Pictorial History*. Norfolk: Domming, 1983.

St. Joseph Catholic Church: Yearbook; 1914-1999. Green Bay, 1999.

Wisconsin Centennial Story of Disaster and Other Unfortunate Events: 1848-1948. Wisconsin State Centennial Committee, 1948, accessed July 2, 2014, http://babel.hathitrust.org/cgi/pt?id=wu.89060458155;view=1up;seq=18.

CITY DIRECTORIES
(All at BCL, listed in chronological order)

Armitage and Pratt's Directory and Business Guide of the City of Green Bay and Borough of Fort Howard for 1872 and 1873. Green Bay: Hoskinson and Follett, 1872.

Dull, J. Alfred. *Green Bay and Fort Howard Directory*. Appleton, WI: Reid and Miller, 1874.

Green Bay, Fort Howard, De Pere and West De Pere Directory for 1881-1882. Green Bay: Louis C. Bold, 1881.

Directory of the Cities of Green Bay, Fort Howard, De Pere and Nicolet, 1884. Milwaukee: A. G. Wright, 1884.

Directory of the Cities of Green Bay, Fort Howard, De Pere and Nicolet, 1888-1887. Milwaukee: A. G. Wright, 1886.

Green Bay and Fort Howard City Directory, 1889-90. Chicago: United States Central, 1889.

Wright's Directory of Green Bay Fort Howard for 1892-93. Milwaukee: A. G. Wright, 1892.

Directory of the Cities of Green Bay and Fort Howard, 1893. Chicago: R. S. Radcliffe, 1893.

Wright's Directory of Green Bay Fort Howard for 1894-1895. Milwaukee: A. G. Wright, 1894.

Wright's Directory of Green Bay for 1896-7. Milwaukee: A. G. Wright, 1896.

Wright's Directory of Green Bay for 1898-9. Milwaukee: A. G. Wright, 1898.

COURT DOCUMENTS

Britton, Appellant, v. The Green Bay and Fort Howard Water Works Company, Respondent, 81 Wis. 48; 51 N. W. 84. Supreme Court of Wisconsin, 1892. Wisconsin Historical Society.

John Atkinson and the Phenix Insurance Company, Plaintiff and Respondent v. The Goodrich Transportation Company, Defendant and Appellant, February 25-March 18, 1884. *Reports of Cases Argued and Determined in the Supreme Court of the State of Wisconsin: Volume LX; February 19-September 23, 1884*. Frederic K. Conover, Official Reporter. Chicago: Callaghan, 1885.

John Atkinson and the Phenix Insurance Company, Plaintiff and Respondent v. The Goodrich Transportation Company, Defendant and Appellant, State of Wisconsin, 60 Wis. 141. *Supreme Court, Cases and Briefs: Volume 178; 1882*. Wisconsin State Law Library, KFW 2400 B7 CB v. 178.

Simmons, James ed. *Reports of Cases Argued and Determined in the Supreme Court of the State of Wisconsin: Volume 69; June 1- November 1, 1887*. Frederic K. Conover, Official Reporter. Chicago: Callaghan, 1888.

MAPS

Baker, J. S. "Map of the City of Green Bay, 1856." Wisconsin Historical Society, accessed November 6, 2014, http://content.wisconsinhistory.org/cdm/singleitem/collection/maps/id/7974/rec/3.

Brauns, A. "Map of the Cities of Green Bay and Fort Howard." Philadelphia: Wm. Brucher, January 1890. City of Green Bay archives.

Foote, C. M. and W. S. Brown, Surveyors and Draughtsmen. "Plat Book of Brown County Wisconsin: Drawn from Actual Surveys and the County Records." Minneapolis: C.M. Foote, 1889. UWGB-ARC.

"Green Bay and Fort Howard: Brown Co. Wisconsin, 1867." Chicago: Chicago Lithography, 1867. Lithograph at BCL.

"Green Bay and Fort Howard, 1893." Milwaukee: C. J. Pauli, 1893. Lithograph at BCL.

National Board of Fire Underwriters, Committee on Fire Prevention. Untitled map of Green Bay showing water supply system, fire stations, and recommended fire apparatus. December, 1915. Green Bay Water Utility archives.

Suydam, J. V. and A. Braun. "Map of the Cities of Green Bay and Fort Howard: From the Late Official Surveys and Map." Green Bay: State Gazette, 1874. Wisconsin Historical Society, accessed November 26, 2014, http://content.wisconsinhistory.org/cdm/singleitem/collection/maps/id/12910/rec/2.

MAPS, SANBORN INSURANCE
(Chronological order)

Green Bay Including Fort Howard. New York: Sanborn, 1879. NPM, 3742/2622.

Green Bay Including Fort Howard. New York: Sanborn, 1883. BCL.

Green Bay Including Fort Howard Wisconsin. New York: Sanborn, 1887. BCL.

Green Bay Including Fort Howard Wisconsin. New York: Sanborn and Perris, 1894. BCL.

Insurance Maps of Green Bay Wisconsin, Brown Co. New York: Sanborn-Perris, 1900. BCL.

Insurance Maps Green Bay, Brown Co., Wisconsin. New York: Sanborn, 1907. BCL.

Insurance Maps of Green Bay, Wisconsin. New York: Sanborn, 1936. BCL.

MISCELLANEOUS

City of Green Bay-Water Department. *Plant Unit Ledger.* Green Bay Water Utility archives.

"Fort Howard Book 1." On Broadway, Inc., historical archive, Green Bay.

"A Gay Day in Green Bay." 1913. NPM. Unedited footage from a documentary film.

"Great Lakes Vessels Online Index." Bowling Green State University, accessed February 12, 2015, http://greatlakes.bgsu.edu/vessel/view/004015.

"Green Bay, 1634-1924." Green Bay: City Commission. City of Green Bay archives.

"Green Bay History: 1800s-1840s." City of Green Bay, accessed January 27, 2014, http://www.ci.green-bay.wi.us/history/1800s.

Green Bay (Wis.) Fire Department. "Log Books and Fire Record Books: 1899-1993." UWGB-ARC, Brown series 176.

"Handtub Database." Handtub Junction, USA, accessed August 16, 2014, http://www.handtubs.com/.

"Historic Broadway: Parts 1 & 2, HB 3-A, HB 4-A." On Broadway, Inc., historical archives, Green Bay.

Howard, Needles, Tammen, and Bergendoff. "City of Green Bay Intensive Resource Survey: Final Report." Milwaukee: 1988.

Kellogg, Louise Phelps. "The Story of Old Fort Howard: An Official Souvenir of the Old Fort Howard; Exhibit At The Wisconsin Tercentennial." Green Bay: Tercentennial, Inc., 1934.

Liedtke, Daniel. "Franklin Hose Company: Interpretive Manual." Heritage Hill, Green Bay, 2008.

Martin, Deborah B. and Sophie Beaumont. "Old Green Bay: Illustrated; 1634-1899." New York: Cheltenham, 1899. City of Green Bay archives.

Peckham, John M. "Complete List of Button Steam Fire Engines." 1976. Waterford Historical Museum, and Cultural Center, Waterford, NY.

Rudolph, Jack. "Franklin Hose Company #3." Heritage Hill, Green Bay. Undated data sheet with introductory and background information.

MUNICIPAL RECORDS

Borough of Fort Howard Clerk. *Record of the Borough Council of the Borough of Fort Howard: Council Records 1856-1865.* UWGB-ARC, Brown County micro series 5, reel 1.

Brown County Clerk. "Assessment Rolls, 1839, 1854, 1859: North Ward Assessment Roll for 1859." UWGB-ARC, Brown series 12, box 1.

Brown County Clerk. "Tax and Land Records: Volumes 1-4, 1839-1867." UWGB-ARC, Brown series 11, box 1.

Brown County Register of Deeds. "Deeds Vol. X: November 1855-September 1856." UWGB-ARC, Brown series 150, vol. 22.

Brown County Treasurer. "Brown County Tax Roll." UWGB-ARC, Brown series 4.

Brown County (Wis.) County Clerk. "Abstracts and Statements of Assessments and Taxes: 1865-1954." UWGB-ARC, Brown series 61, box 1.

Charter and Consolidated Ordinances of the City of Green Bay with Amendments Thereto. Green Bay: Rummel and Bender, 1882. UWGB-ARC.

Charter of the City of Green Bay: All General Ordinances Passed Since the Organization of the City; To April 1st, 1872. F. Burkard, 1872. UWGB-ARC.

City of Fort Howard, City Clerk. *Proceedings of the Common Council of Fort Howard: 1865-1895.* UWGB-ARC, Brown series 35.

City of Fort Howard, City Clerk. *Ordinances of the City of Fort Howard: 1868-1894.* UWGB-ARC, Brown series 36.

City of Green Bay, City Clerk. *Records of the Borough of Green Bay: Proceedings of the Common Council; 1853-.* UWGB-ARC, Brown series 34, vol. 1-8.

City of Green Bay, Clerk. *City of Green Bay Ordinances: 1854-1909.* UWGB-ARC, Brown series 57, vol. 1-3.

Journal of Proceedings of the County Board of Equalization, Brown County: Clerk, Equalization Board Proceedings and Tax Record; 1858-1859. UWGB-ARC, Brown small series 6, vol. 1.

"Miscellaneous Brown County Financial and Legal Documents: Folder 3; Exempt Property Statements 1870-September 24, 1900." NPM, MSS2003.25.

Wisconsin Secretary of State. "Tax-Exempt Property Record: 1872-1874." Wisconsin Historical Society, series 229.

NEWSPAPERS

Appleton Crescent

Appleton Post-Crescent

Bay City Press

Brown County Democrat

Brown County Herald

Daily Milwaukee News

Daily State Gazette

De Pere News

Fort Howard Herald

Fort Howard Journal

Fort Howard Monitor

Fort Howard Review

Green Bay Advocate

Green Bay Globe

Green Bay Gazette

Green Bay Intelligencer

Green Bay Press-Gazette

Green Bay Republican

Green Bay Semi-Weekly Gazette

Green Bay Spectator

Green Bay State Gazette

Green Bay Weekly Gazette

Janesville Gazette

Madison Wisconsin State Journal

Milwaukee Sentinel

New York Times

Semi-Weekly Wisconsin

State Gazette

Stevens Point Journal

Twin City Index

Wisconsin Democrat

Wisconsin Free Press

ORAL HISTORIES (NPM)

Delwiche, August. November 4, 1936, 711 Porlier Street, Green Bay.

Faulkner, Henry. October 30, 1936, 120 South Chestnut Street, Green Bay.

Herrmann, Carl. November 10, 1936, 223 Main Street, Green Bay.

Lison, Watson. October 27, 1936, 148 North Oakland Avenue, Green Bay.

Tickler, Albert. October 3, 1936, 939 South Quincy Street, Green Bay.

OTHER GOVERNMENT RECORDS

"An Act to Incorporate the Borough of Fort Howard." Wisconsin Chapter 535, Private and Local Laws 1856, Section 15, Sub Section 9, 1210-1. State of Wisconsin Legislative Reference Bureau.

"Green Bay City Charter." Laws of Wisconsin, 1836-1838, Chapter 66, Section 8, 378. State of Wisconsin Legislative Reference Bureau.

Office of the Quartermaster General. "Consolidated Correspondence File, 1794-1915: Fort Howard, Wisconsin." National Archives and Records Administration, Washington, D.C., record group 92, box 857.

Private and Local Laws Passed by the Legislature of Wisconsin in the Year 1868. Madison: Atwood and Rubles. UWGB-ARC.

Wisconsin State Law, Laws of 1870, Chapter 56, Section 34, 98, State of Wisconsin Legislative Reference Bureau.

PERIODICALS

Liedtke, Daniel and Tony LaLuzerne. "A Brief History of Early Firefighting in Green Bay." *The Historical Bulletin* 28, no. 3 (September 2010).

Neville, Arthur C. "Early Ferries and Bridges Across The Fox River." *Green Bay Historical Bulletin* 3, no. 1 (January/February 1927).

Parker, Barton L. "The History and Location of Fort Howard." *Green Bay Historical Bulletin* 5, no. 4 (October, November, December 1929).

Telzrow, Michael. "Bonds of Brotherhood: Green Bay's Volunteer Firefighters." *Voyageur* 27, no. 2 (Winter/Spring 2011), 48.

SPECTATOR INSURANCE BOOKS

The Spectator Insurance Books are a series of extensive, in-depth reports prepared for the insurance industry and include detailed information on fire departments and fire protection. For this project, Spectator books utilized were from 1873 and annually from 1879 through 1916. All were found through: http://babel.hathitrust.org, http://books.google.com, or https://archive.org.

VOLUNTEER FIRE COMPANY RECORD BOOKS (NPM)

Guardian Fire Co.'s, Roll Call. Item no. 759/4479 MS4 A4.

Notebook, Franklin Fire Hose Co. No.3. Item no. 3485/1312.

Record Book, Guardian Engine Co. No. 2. Item no. MS4C1, box 4, 762/4482.

Secretary's Roll, Guardian Fire Co., Green Bay. Item no. MS4A4, 761/4481.

Treasurer's Report, Guardian Fire Co., Green Bay. Item no. MS4A4, 760/4480.

Washington Hook and Ladder Co., Log Book. Item no. MS4A4, 1999. 19.3.

Endnotes

ABBREVIATIONS

BCL: Brown County Library, Local History and Genealogy Department, Green Bay, Wisconsin

GBFD: Green Bay Fire Department

NARA: Office of the Quartermaster General, "Consolidated Correspondence File, 1794-1915: Fort Howard, Wisconsin," National Archives and Records Administration, Washington, D.C., Record group 92, box 857.

NPM: Neville Public Museum of Brown County, Green Bay, Wisconsin

UWGB-ARC: University of Wisconsin Green Bay-Area Research Center, Cofrin Library, Green Bay, Wisconsin

KEY

Shortened citations are defined at each bullet point for the following frequently used municipal records.

FORT HOWARD COMMON COUNCIL MEETING MINUTES

- Borough of Fort Howard Clerk, *Record of the Borough Council of the Borough of Fort Howard: Council Records 1856-1865*, UWGB-ARC, Brown County micro series 5, reel 1 (hereafter cited as Fort Howard Common Council, 1856-65).

City of Fort Howard, City Clerk, *Proceedings of the Common Council of Fort Howard: 1865-1895*, UWGB-ARC, Brown series 35:

- April 18, 1865-April 17, box no. 1 (hereafter cited as Fort Howard Common Council, 1865-76).

- May 1, 1876-December 21, 1885, box no. 2 (hereafter cited as Fort Howard Common Council, 1876-85).

- January 4, 1886-June 18, 1894, box no. 3 (hereafter cited as Fort Howard Common Council, 1886-94).

- July 6, 1894-April 9, 1895, box no. 4 (hereafter cited as Fort Howard Common Council, 1894-95).

FORT HOWARD ORDINANCES

City of Fort Howard, City Clerk, *Ordinances of the City of Fort Howard: 1868-1894*, UWGB-ARC, Brown series 36, box 1:

- 1868-1886, vol. 1 (hereafter cited as Fort Howard Ordinances, 1868-86).

- Aug. 18, 1886-Aug. 24, 1894, vol. 2 (hereafter cited as Fort Howard Ordinances, 1886-94).

GREEN BAY COMMON COUNCIL MEETING MINUTES

City of Green Bay, City Clerk, *Records of the Borough of Green Bay: Proceedings of the Common Council*; 1853-, UWGB-ARC, Brown series 34:

- November 9, 1853-December 3, 1859, box 1, vol. 1 (hereafter cited as Green Bay Common Council, 1853-59).

- December 5, 1859-April 9, 1870, box 1, vol. 2 (hereafter cited as Green Bay Common Council, 1859-70).

- April 12, 1870-December 17, 1875, box 2, vol. 3 (hereafter cited as Green Bay Common Council, 1870-75).

- December 20, 1875-December 9, 1879, vol. 4 (hereafter cited as Green Bay Common Council, 1875-79).

- January 2, 1880-July 11, 1883, vol. 5 (hereafter cited as Green Bay Common Council, 1880-83).

- July 20, 1883-July 6, 1886, 1883-86 vol. 6 (hereafter cited as Green Bay Common Council, 1883-86).

- July 19, 1886-February 5, 1892, vol. 7 (hereafter cited as Green Bay Common Council, 1886-92).

- April 12, 1892-February 19, 1897, vol. 8 (hereafter cited as Green Bay Common Council, 1892-97).

GREEN BAY ORDINANCES

City of Green Bay, Clerk, *City of Green Bay Ordinances: 1854-1909*, UWGB-ARC, Brown series 57:

- May 6, 1854-June 25, 1875, vol. 1 (hereafter cited as Green Bay Ordinances, 1854-75).
- June 25, 1875-October 10, 1892, vol. 2 (hereafter cited as Green Bay Ordinances, 1875-92).
- June 9, 1893-July 1, 1903, vol. 3 (hereafter cited as Green Bay Ordinances, 1893-1903).

CHAPTER 2
FALTERED BEGINNINGS

1. *Wisconsin Free Press*, January 9, 1836.
2. Green Bay City Charter, Laws of Wisconsin, 1836-1838, Chapter 66, Section 8, 378, State of Wisconsin Legislative Reference Bureau.
3. *Wisconsin Democrat*, June 25, 1839. Original ordinance; *Green Bay Republican*, January 8, 1842. Names and numbers on buckets.
4. *Wisconsin Democrat*, June 25, 1839. Original ordinance.
5. *Green Bay Republican*, January 1, 1842. Refers to a large fire one year earlier.
6. Bella French ed., *The American Sketch Book: History of Brown County, Wisconsin* (Green Bay: American Sketch Book, 1876), 3:104-107.
7. *Green Bay Republican*, December 25, 1841. Date of fire and damage; Ephraim Shaler to Quartermaster General Thomas Jesup, January 14, 1842, NARA. Describes an entire block destroyed.
8. Shaler, January 14, 1842.
9. Kenneth Holcomb Dunshee, *Enjine!-Enjine! A Story of Fire Protection* (New York: Harold Vincent Smith for Home Insurance, 1939), 18.
10. Dunshee, *Enjine!-Enjine!*, 13.
11. *Green Bay Republican*, January 1, 1842.
12. Henry S. Baird to Ephraim Shaler, December 29, 1841, NARA.
13. Shaler, January 14, 1842. Notes attached to this letter, dated February 9 and February 22, 1842.
14. Ephraim Shaler to Quartermaster General Thomas Jesup, September 7, 1842, NARA.
15. Ibid. Note dated September 22, 1842, attached to September 7, 1842 letter.
16. Ephraim Shaler to Quartermaster General Thomas Jesup, October 31, 1842, NARA.
17. Daniel Whitney to Secretary of War J. C. Spencer, November 23, 1842, NARA.
18. *Green Bay Advocate*, February 11, 1847.
19. *Green Bay Advocate*, February 11, 1847.
20. Daily State Gazette, September 18, 1891.
21. Hand Engine Makers (American), Handtub Junction USA, accessed August 16, 2014, http://www.handtubs.com/.
22. *Green Bay Republican*, January 1 and January 8, 1842.
23. *Green Bay Republican*, January 1, 1842.
24. *Green Bay Republican*, April 16, 1842.
25. Daniel Whitney to Secretary of War John C. Spencer, November 23, 1842, NARA.
26. *Green Bay Advocate*, February 18, 1847.
27. *Green Bay Advocate*, April 13, 1848.
28. *Green Bay Advocate*, February 25, April 1, and July 1, 1847.
29. *Green Bay Advocate*, September 6, 1849.
30. *Green Bay Advocate*, March 21, 1850.
31. *Green Bay Advocate*, March 21, 1850.
32. J. B. Collins to Quartermaster General Thomas Jesup, October 7, 1849, NARA; Francis Lee to E. Backus, January 11, 1850, NARA.
33. *Green Bay Advocate*, March 28, 1850.
34. *Green Bay Advocate*, March 28, 1850.

35. *Green Bay Advocate*, February 13, 1851.

36. *Green Bay Advocate,* May 15, 1851, Fire engine arrives; Green Bay Common Council, 1853-59, 1. The November 1853 fire destroyed contemporaneous official records from May, 1851 including purchase of the Smith hand pumper; ibid., 3. Common Council wrote James Smith in 1853 for balance on hand pumper sale; ibid., 164. Minutes from 1856 clearly state that Green Bay bought a Smith hand pumper in 1851.

37. *Green Bay Advocate*, May 15, 1851.

38. *Green Bay Advocate,* October 2, 1851. Name, numbers, and parade quote.

39. *Daily State Gazette*, August 21, 1875; Green Bay Common Council, 1870-75, 636.

40. *Green Bay Advocate*, March 4, 1852. First quote; *Green Bay Advocate*, April 1, 1852. Second quote.

41. *Green Bay Spectator*, August 24, 1852.

42. *Green Bay Advocate*, April 7, 1853.

43. *Green Bay Advocate*, July 7, 1853.

CHAPTER 3
DAWN OF THE GREEN BAY FIRE DEPARTMENT

1. Population summary sheet, BCL, undated.

2. *Appleton Crescent*, November 5, 1853.

3. *Milwaukee Sentinel*, November 4, 1853. Damage estimate cost; *New York Times*, November 8 and November 9, 1853; Green Bay Common Council, 1853-59, 1. All previous common council minutes lost in fire.

4. Jack Rudolph, *Green Bay Press-Gazette*, February 27, 1960.

5. *Appleton Crescent*, November 12, 1853. Man missing; *Green Bay Advocate*, April 27, 1854. Body found.

6. *Green Bay Advocate*, December 15, 1853. Quotes and extensive description of fire.

7. *Green Bay Advocate*, December 15, 1853.

8. Green Bay Common Council, 1853-59, 1. Bill to repair hand pumper; ibid., 55. City marshal directed to put fire engine in order.

9. *Green Bay Advocate*, November 30, 1854. Extensive description of fire.

10. *Green Bay Advocate,* December 7 and December 14, 1854.

11. Green Bay Common Council, 1853-59, 63.

12. Green Bay Common Council, 1853-59, 64.

13. French, *American Sketch Book*, 147. Organizational meeting date; *Green Bay Advocate,* December 21, 1854. Fire company name; *Green Bay Advocate,* August 28, 1856. All members of German birth.

14. *Green Bay Advocate,* December 21, 1854.

15. *Green Bay Advocate,* January 4, 1855.

16. French, *American Sketch Book*, 147. Rough and Ready nickname.

17. *Green Bay Advocate*, June 1, 1854. Old engine house repaired; *Green Bay Advocate,* October 6, 1853. Town house located on Adams Street.

18. Green Bay Common Council, 1853-59, 129, and 136. Bill paid with substantial interest.

19. Brown County Clerk, "Assessment Rolls, 1839, 1854, 1859: Navarino and Astor, 1882," UWGB-ARC, Brown series 12, box 1, vol. 3.

20. *Green Bay Advocate,* February 15, 1855.

21. *Green Bay Advocate,* January 4, 1855.

22. Green Bay Common Council, 1853-59, 71.

23. Green Bay Common Council, 1853-59, 93. Hose purchase; ibid., 1, 118, and 188. Repairs on fire engine.

24. *Green Bay Advocate,* August 30, 1855.

25. *Green Bay Advocate,* August 21 and August 28, 1856.

26. *Green Bay Advocate,* January 2, 1856, January 8, 1857 and January 7, 1858.

27. *Green Bay Advocate,* February 28, 1856. Quotes.

28. *Green Bay Advocate,* May 8, 1856.

29. *Green Bay Advocate,* February 28, 1856.

30. Green Bay Common Council, 1853-59, 190. Petition presented; ibid., 194.

31. Green Bay Common Council, 1853-59, 199.

32. *Green Bay Advocate,* March 17, 1859; *Record Book, Guardian Engine Company No. 2,* May 5, 1890, NPM, MS4C1, box 4, 762/4482 (hereafter cited as Guardian Record Book). Meeting minutes uses Wide Awake nickname.

33. Green Bay Common Council, 1853-59, 211.

34. *Green Bay Advocate,* April 22, 1858. New fire engine arrives; Green Bay Common Council, 1853-59, 421. Bill from Button and Blake; ibid., 439. Direct identification of No. 2 engine as Button and Blake.

35. *Green Bay Advocate,* April 22, 1858.

36. *Green Bay Advocate,* July 8 and July 15, 1858.

37. *Treasurer's Report, Guardian Fire Co., Green Bay,* Roll Call, 1859, NPM, MS4A4, 760/4480 (hereafter cited as Guardian Treasurer's Report); *Secretary's Roll, Guardian Fire Co., Green Bay,* Roll Call, 1861, NPM, MS4A4, 761/4481 (hereafter cited as Guardian Secretary's Roll).

38. Green Bay Common Council, 1853-59, 349.

39. Green Bay Common Council, 1853-59, 403-4. Apparatus accepted; *Green Bay Advocate,* August 5, 1848. Weise as wagon maker; *Green Bay Advocate,* November 11, 1858. Weise as Germania No. 1 member.

40. Green Bay Common Council, 1853-59, 404.

41. Green Bay Common Council, 1853-59, 414.

42. Green Bay Common Council, 1853-59, 434.

43. Brown County Clerk, "Assessment Rolls, 1839, 1854, 1859: North Ward Assessment Roll for 1859," UWGB-ARC, Brown series 12, box 1.

44. Green Bay Common Council, 1853-59, 462.

45. Green Bay Common Council, 1853-59, 412 and 420. Numerous common council motions; Green Bay Ordinances, 1854-75, Ordinance No. 45, March 27, 1858, 80; ibid., Ordinance No. 48, April 24, 1858, 86.

46. *Green Bay Advocate,* April 15, 1858.

47. Green Bay Common Council, 1853-59, 556.

48. Green Bay Ordinances, 1854-75, Ordinance No. 76, October 8, 1859, 130.

49. *Green Bay Advocate,* April 26, 1860.

50. *Green Bay Advocate,* May 3, 1860.

51. *Green Bay Advocate,* May 2, 1861.

52. *Green Bay Advocate,* May 15, 1863.

53. *Green Bay Advocate,* April 26, 1860. Franklin using Old Croc; Green Bay Common Council, 1859-70, 16 and 18.

54. Green Bay Common Council, 1853-59, 547.

55. *Green Bay Advocate,* September 27, 1860.

56. *Green Bay Advocate,* April 26, 1860.

57. Green Bay Common Council, 1853-59, 434.

58. Green Bay Common Council, 1853-59, 551. Part of lot. 114 purchased for Engine House No. 2; Green Bay Common Council, 1859-70, 4. New Engine House No. 2 accepted.

59. *Green Bay Advocate,* June 14, 1860; *Green Bay, Fort Howard, De Pere and West De Pere Directory for 1881-1882* (Green Bay: Louis C. Bold, 1881), 18, BCL; *Green Bay Including Fort Howard* (New York: Sanborn, October 1879), 3, NPM, 3742/2622 (hereafter cited as Sanborn, 1879). Insurance map shows Cherry Street location.

60. Green Bay Common Council, 1859-70, 37, 43, and 45.

61. *Green Bay Advocate,* June 21, 1860.

62. Green Bay Common Council, 1853-59, 70.

63. Green Bay Common Council, 1859-70, 59, 115, and 117.

64. Green Bay Common Council, 1859-70, 164 and 213.

65. Green Bay Common Council, 1859-70, 116.

66. Green Bay Common Council, 1853-59, 512.

67. Green Bay Common Council, 1853-59, 514.

68. Green Bay Common Council, 1859-70, 64, 66, 68, 74, 218, and 244. Small bills ($2-$10) paid, mostly during winter; *Green Bay Advocate,* January 31, 1861. Difficulty responding to house fire through deep snow.

69. Green Bay Common Council, 1853-59, 174.

70. Green Bay Common Council, 1859-70, 194.

71. Green Bay Common Council, 1859-1870, 83.

72. *Green Bay Advocate,* January 2, 1856.

73. *Green Bay Advocate,* December 11, 1856. First Party; *Green Bay Advocate,* February 18, 1858. Second party.

74. *Green Bay Advocate,* June 16, 1859; Jack Rudolph, *Green Bay Press-Gazette,* October 8, 1960; *Green Bay Advocate,* November 12, 1863. Klaus Hall destroyed by fire.

75. Green Bay Ordinances, 1854-75, Ordinance No. XVIII, November 25, 1854, 40.

76. Green Bay Ordinances, 1854-75, Ordinance No. 61, November 16, 1858, 110.

77. Green Bay Common Council, 1853-59. Multiple instances when common council assigned or replaced fire wardens.

78. *Green Bay Advocate,* October 27, 1859.

79. *Green Bay Advocate,* April 10, 1856.

80. Green Bay Ordinances, 1854-75, Ordinance No. 70, November 1, 1859, 119.

81. Green Bay Ordinances, 1854-75, Ordinance No. 70, November 1, 1859, 119.

82. Green Bay Ordinances, 1854-75, Ordinance No. 70, November 1, 1859, 119.

83. Green Bay Ordinances, 1854-75, Ordinance No. 99, January 15, 1864.

84. *Green Bay Advocate,* August 7, 1856.

85. *Green Bay Advocate,* March 19, 1857.

86. *Green Bay Advocate,* July 9, 1857

87. *Green Bay Advocate,* October 21, 1858.

88. *Green Bay Advocate,* February 23, 1860.

89. *Green Bay Advocate,* February 23, 1860.

90. *Milwaukee Sentinel,* February 25, 1860.

91. *Green Bay Advocate,* April 19, 1860.

92. *Green Bay Advocate,* May 3, 1860.

93. *Green Bay Advocate,* July 12, 1860.

94. *Bay City Press,* April 5, 1862.

CHAPTER 4
TRANSITION TO STEAM FIRE ENGINES

1. *Green Bay Advocate,* May 15, 1863.

2. Population summary sheet, BCL, undated.

3. Green Bay Ordinances, 1854-75, Ordinance No. XVIII, November 25, 1854, 40; ibid., Ordinance No. 48, April 24, 1858, 86.

4. *Daily Milwaukee News,* November 12, 1863. This paper states the fire started at 3 am, but 2 am is accurate per later Green Bay papers.

5. Green Bay Ordinances, 1854-75, Ordinance No. 70, November 1, 1859, 119.

6. *Green Bay Advocate,* November 12, 1863.

7. *Green Bay Advocate,* November 12, 1863.

8. *Green Bay Advocate,* November 12, 1863.

9. *Green Bay Advocate,* November 12, 1863.

10. Green Bay Ordinances, 1854-75, Ordinance No. 70, November 1, 1859, 119.

11. *Green Bay Advocate,* November 12, 1863. Very detailed description of the fire and aftermath.

12. French, *American Sketch Book,* 168.

13. *Green Bay Advocate,* November 12, 1863.

14. *Green Bay Advocate,* December 3, 1863.

15. *Green Bay Advocate,* February 4, 1864.

16. *Green Bay Advocate,* June 9, 1864.

17. *Green Bay Advocate,* April 24, 1862.

18. Green Bay Ordinances, 1854-75, Ordinance No. 45, March 27, 1858, 80.

19. *Green Bay Gazette,* August 11, 1866; *Green Bay Advocate,* August 9 and August 16, 1866.

20. W. Fred Conway, *Those Magnificent Old Steam Fire Engines* (New Albany, IN: Fire Bluff House, 1997), 13-16.

21. Green Bay Common Council, 1880-83, 301; *Green Bay Advocate*, March 25, 1880; *Green Bay Advocate*, November 27, 1884.

22. *Green Bay Gazette*, April 14, 1866.

23. Green Bay Common Council, 1859-70, 355.

24. *Green Bay Advocate*, February 28, 1867.

25. *Green Bay Advocate*, August 9, 1866.

26. *Green Bay Advocate*, October 26, 1866.

27. Green Bay Common Council, 1859-70, 402.

28. Green Bay Common Council, 1859-70, 404.

29. Green Bay Common Council, 1859-70, 436; *Green Bay Gazette*, December 28, 1867.

30. Green Bay Common Council, 1859-70, 437.

31. Green Bay Common Council, 1859-70, 461; Green Bay Ordinances, 1854-75, Ordinance No. 152, May 25, 1868.

32. *Green Bay Gazette*, March 21, 1868.

33. *Green Bay Gazette*, April 14, 1866, Mayor Robinson; *Green Bay Gazette*, April 18, 1868. Mayor Klaus.

34. *Green Bay Advocate*, October 29, 1868; *Green Bay Gazette*, October 31, 1868.

35. J. Alfred Dull, *Green Bay and Fort Howard Directory* (Appleton, WI: Reid and Miller, 1874), 38-9, BCL (hereafter cited as City Directory, 1874). Enterprise No. 1 steamer described as second class; Conway, *Steam Fire Engines*, 277. Second class description; ibid., 41. Pump flow-rate as gallons per minute.

36. Green Bay Common Council, 1859-70, 569. Steamer formal name; Green Bay Common Council, 1859-70, 505 and 510. New fire company name; *Green Bay Advocate*, September 17, 1868. New fire company name; *Green Bay Advocate*, October 7, 1875. Fire company motto on caps.

37. Green Bay Common Council, 1859-70, 437, 513, and 569; *Green Bay Advocate*, November 5, 1868.

38. *Green Bay Advocate*, November 5, 1868.

39. *Green Bay Advocate*, November 26, 1868.

40. Green Bay Common Council, 1859-70, 494.

41. Green Bay Common Council, 1859-70, 495.

42. City Directory, 1874, 49 and 68.

43. Green Bay Common Council, 1859-70, 588. 1869 salary; Green Bay Common Council, 1870-75, 89. 1871 salary; ibid., 274. 1873 salary; Green Bay Common Council, 1875-79, 259, 1877. Salaries for engineers, No. 1 $800 and No. 2, $600; Green Bay Common Council, 1880-83, 40, 171, 279, and 404. Early-1880s salaries; Green Bay Common Council, 1883-86, 80, 180, and 266. Mid-1880s salaries.

44. Conway, *Steam Fire Engines*, 76. Steamer dimensions.

45. Green Bay Common Council, 1859-70, 493. Contract to move old engine house; ibid., 503. Engine house lot on South Washington now empty; *Green Bay Gazette*, September 19, 1868. Old engine house moved to south ward; *Green Bay Advocate*, October 8, 1868. Old engine house at triangle park; Carl Herrmann, oral history, November 10, 1936, 223 Main Street, Green Bay, NPM. Old engine house at triangle park.

46. Sanborn, 1879, 3. Station profile and tower height; *Green Bay Advocate*, October 1, 1868, October 29, 1868, and January 7, 1869, *Green Bay State Gazette*, January 9, 1869.

47. Green Bay Common Council, 1859-70, 510.

48. *Green Bay Advocate*, June 25, 1868. Plan to use old Engine House No. 1 as another engine house in south ward; Jack Rudolph, "Fire Fightin' Firemen," undated manuscript, GBFD historical archives. Smith hand pumper from Germania No. 1 to Astor Fire Company.

49. *Green Bay Advocate*, March 19, 1868.

50. Green Bay Common Council, 1859-70, 437.

51. *Green Bay Gazette*, April 18, 1868. Mayor's address; Green Bay Common Council, 1859-70, 437 and 440. Cistern construction cost part of bond.

52. Green Bay Common Council, 1859-70, 503. Volume of cistern calculated from dimensions recorded in common council minutes.

53. Conway, *Steam Fire Engines*, 76.

54. Green Bay Common Council, 1859-70, 508.

55. *Green Bay Advocate*, November 5, 1868.

56. Green Bay Common Council, 1859-70; Green Bay Common Council, 1870-75. Numerous examples in both references.

57. *Green Bay Advocate*, October 14, 1869.

58. Green Bay Common Council, 1870-75, 187 and 190.

59. Green Bay Ordinances, 1854-75, Ordinance No. 168, November 20, 1968.

60. Green Bay Common Council, 1859-70, 589; *Green Bay Advocate*, October 28, 1869. Bell weight.

61. Green Bay Common Council, 1859-70, 607.

62. Green Bay Common Council, 1870-75, 369. Bell recast; ibid., 378 and 391. Tower rebuilt.

63. Ed Schuch, interview by author, October 14, 2014, 1421 Coral Court, Green Bay; *St. Joseph Catholic Church: Yearbook; 1914-1999* (Green Bay: 1999).

64. *Green Bay Advocate*, February 26, 1874.

65. *Green Bay Advocate*, May 13, 1875. Signal system; *Green Bay Advocate*, June 3, 1875. Open tower, better sound.

66. Green Bay Common Council, 1870-75, 90.

67. Green Bay Common Council, 195. Petition; *Green Bay Advocate*, June 6, 1872. Petition; Green Bay Common Council, 1870-75, 202. Clapp and Jones contract.

68. *Green Bay Advocate*, August 29, 1872.

69. Green Bay Common Council, 1870-75, 202.

70. *State Gazette*, August 24, 1872.

71. *Green Bay Advocate*, August 29, 1872.

72. Green Bay Common Council, 1870-75, 208.

73. Green Bay Common Council, 1870-75, 508. Hose cart proposed; ibid., 545. Hose cart accepted.

74. Green Bay Common Council, 1870-75, 212.

75. Green Bay Common Council, 1870-75, 51 and 57; *Green Bay Advocate*, December 15, 1870.

76. *Green Bay Advocate*, April 13, 1871, May 18, 1871, August 1, 1872, and March 6, 1873.

77. Green Bay Common Council, 1870-75, 218. 1872 cisterns; ibid., 447. Roosevelt/Walnut cistern, 1874; ibid., 463. Irwin/Walnut cistern, 1874.

78. *Green Bay Advocate*, January 7, 1875.

79. *Green Bay Advocate*, September 3, 1874. Water volume used reported as 1,500 barrels, converted to gallons (41.5 gallons/barrel).

80. Green Bay Common Council, 1870-75, 436.

81. Green Bay Common Council, 1859-70, 164. Night watchman duties in 1862; Green Bay Ordinances, 1854-75, Ordinance No. 14, New Series, March 1, 1872; Green Bay Common Council, 1875-79, 235. Water hole maintenance duties in 1876.

82. Green Bay Common Council, 1875-79, 426.

83. *Green Bay Advocate*, January 20, 1881.

84. Green Bay Common Council, 1870-75, 674.

85. *Green Bay Advocate*, February 11, 1875.

86. Guardian Record Book, March 23, 1875.

87. Guardian Record Book, March 23, 1875. Drivers staying overnight; ibid., May 10, 1875. Beds purchased.

88. Watson Lison, oral history, October 27, 1936, 148 North Oakland Avenue, Green Bay, NPM.

89. Guardian Record Book, March 23, 1875 and October 11, 1875. Policy set and report of money from team.

90. Guardian Record Book, January 8, 1877.

91. *Green Bay Advocate*, March 29, 1877.

92. *Green Bay Advocate*, September 30, 1875.

93. *Green Bay Advocate*, January 13, 1876.

94. *Green Bay Advocate*, October 12, 1876.

95. *Green Bay Advocate*, September 30, 1875.

96. *Green Bay Advocate*, October 17, 1878.

97. *Green Bay Advocate*, September 22, 1881.

98. August Delwiche, oral history, November 4, 1936, 711 Porlier Street, Green Bay, NPM.

99. Green Bay Common Council, 1883-86, 19; *Green Bay Advocate*, September 27, 1883.

100. Green Bay Common Council, 1883-86, 43-7 and 54-5; Fort Howard Common Council, 1876-85, 547-8.

101. *Green Bay Advocate*, April 20, 1876. Broom on Engine House No. 1 masthead; Albert Tickler, oral history, October 3, 1936, 939 South Quincy Street, Green Bay, NPM. Broom on steamer.

102. Green Bay Common Council, 1880-83, 217.

103. Green Bay Common Council, 1870-75, 226; *Green Bay Advocate*, October 31, 1872.

104. *Green Bay Advocate*, August 9, 1877.

105. Green Bay Common Council, 1870-75, 313, 380, and 384.

106. Green Bay Common Council, 1883-86, 209.

CHAPTER 5
CONSOLIDATION AND PROGRESS

1. Green Bay Common Council, 1870-75, 202.

2. *Green Bay Advocate,* July 1, 1875.

3. Green Bay Common Council, 1870-75, 638; *Daily State Gazette*, August 21, 1875.

4. Green Bay Common Council, 1870-75, 638.

5. *Daily State Gazette*, August 21, 1875.

6. Green Bay Common Council, 1880-83, 257.

7. Green Bay Common Council, 1870-75, 636; *Daily State Gazette*, August 21, 1875.

8. *Green Bay Advocate,* July 1, 1875.

9. Green Bay Common Council, 1870-75, 492; *Green Bay Advocate*, December 31, 1874.

10. Green Bay Common Council, 1875-79, 498. Old Croc used by local company; *Daily State Gazette*, September 18, 1891. Stored behind Engine House No. 2.

11. *A Gay Day in Green Bay*, 1913, NPM. Unedited footage from a documentary film.

12. Green Bay Common Council, 1870-75, 508 and 545.

13. Green Bay Common Council, 1875-79, 94, 108, 114, and 146.

14. Green Bay Common Council, 1875-79, 583.

15. *Green Bay Advocate,* September 14, 1876.

16. Green Bay Common Council, 1875-79, 192.

17. Green Bay Common Council, 1875-79, 102. Chief engineer; ibid., 478. Assistant chief engineer.

18. Green Bay Common Council, 1875-79, 102 and 478; Green Bay Common Council, 1883-86, 92.

19. Green Bay Ordinances, 1875-1892, Chapter VIII, October 4, 1875, 60.

20. Green Bay Common Council, 1875-79, 478 and 574; Green Bay Common Council, 1880-83, 131.

21. City Directory, 1874. City directory lists 120 firefighters not including Washington No. 1. Added twenty firefighters as conservative estimate of Washington membership to give 140 firefighters total in Green Bay.

22. *Green Bay Advocate, January* 7, 1875. Article states 110 firefighters, not including Guardian No. 2; Guardian Record Book. Guardian No. 2 has twenty-one members for total of 131 GBFD firefighters in early 1875; *Green Bay Advocate,* March 29, 1877. GBFD firefighters in 1877.

23. *The Insurance Year Book for 1879-80. Carefully Corrected to June 1, 1880.* (New York, Chicago: Spectator, 1880), 290, http://babel.hathitrust.org/cgi/pt?id=wu.89095809067;view=1up;seq=7.

24. *Charter and Consolidated Ordinances of the City of Green Bay with Amendments Thereto* (Green Bay: Rummel and Bender, 1882), 41, UWGB-ARC.

25. The Insurance Year Book. 1883-84. Carefully Corrected to June 20, 1883. (New York: Spectator, 1883), 363, http://babel.hathitrust.org/cgi/pt?id=nyp.33433082318662;view=1up;seq=11.

26. Green Bay Common Council, 1883-86, 64; Green Bay Advocate, January 17, 1884. Fire on Klaus block.

27. *Green Bay Advocate*, February 12, 1885. Chief engineer report.

28. *Green Bay Advocate,* February 12, 1885; Green Bay Common Council, 1883-86, 250; Green Bay Common Council, 1886-92, 45.

29. *Green Bay and Fort Howard City Directory, 1889-90* (Chicago: United States Central, 1889), 34, BCL (hereafter cited as City Directory, 1889-90).

30. *Wright's Directory of Green Bay Fort Howard for 1892-93* (Milwaukee: A. G. Wright, 1892), 27, BCL.

31. Green Bay Common Council, 1875-79, 439.

32. Green Bay Common Council, 1875-79, 439. Quote; ibid., 444. Common council authorizes disbursements to fire companies; *Washington Hook and Ladder Co., Log Book*, February 1885-January 1892, NPM, MS4 A4, 1999. 19.3 (hereafter cited as Washington Log Book). Minutes repeatedly record $45 from the city first of each year; *Notebook, Franklin Fire Hose Co. No.3*, NPM, 3485/1312, January 1, 1889 (hereafter cited as Franklin Notebook). City gives $45 to Franklin.

33. Green Bay Common Council, 1875-79, 691; Green Bay Common Council, 1880-83, 234 and 365; Green Bay Common Council, 1886-1892, 142; Green Bay Common Council, 1892-1897, 21, 29, 36, 42, 49, 57, and 66.

34. *Green Bay Advocate*, April 20, 1876; Wisconsin State Law, Laws of 1870, Chapter 56, Section 34, 98, State of Wisconsin Legislative Reference Bureau.

35. *Green Bay Advocate*, March 13, 1873.

36. *Green Bay Advocate*, March 18, 1875; *Green Bay Advocate,* March 6, 1879.

37. *Green Bay Advocate*, February 9, 1882; *Green Bay Advocate,* February 23, 1883; Washington Log Book. Financial summary listed every January 1 for 1888 to 1890, includes about $200 from insurance premium payouts. Presume that Germania No. 1 and Guardian No. 2 received similar amounts. Franklin Notebook, undated meeting, early 1890. Franklin received $62.45 from insurance premiums, suggesting this company received less than others.

38. *Daily State Gazette*, February 8, 1892.

39. Guardian Treasurer's Report. Numerous instances where treasurer reported amount of dues collected at meetings.

40. *Guardian Fire Co.'s, Roll Call*, NPM, 759/4479 MS4 A4. Constitution and By-Laws, insert with front cover.

41. Franklin Notebook, August 8, 1887 as well as Constitution and By-Laws.

42. Guardian Treasurer's Report. Numerous entries in 1873 show 25 cent fines for missing meetings; Washington Log Book. Numerous meeting records show 10 cents collected from each member; Franklin Notebook, Constitution and By-Laws.

43. Washington Log Book. Undated list of fines for various infractions under Treasurer's Notes at back of book; Guardian Treasurer's Report. Numerous entries in 1873 show 25 or 50 cent fines for missing fire alarms.

44. Franklin Notebook, June 4, 1889. Fine for failing to clean engine house; Guardian Treasurer's Report. Entries in June/July 1873 show $2.50 fine for missing inspection day.

45. Washington Log Book, February 5, 1888. One member fined for this infraction, under Treasurer's Notes.

46. Guardian Treasurer's Report. Based on tally of fine and dues collected from members during mid-1870s.

47. Franklin Notebook. Tally of fines and dues collected at most meetings in 1890-91.

48. Washington Log Book. Tally of yearly fines/dues under Treasurer's Notes.

49. Washington Log Book. Under Treasurer's Notes, Fenton Fox expelled for excessive fines; Guardian Record Book. Expelled members: four on July 13, 1874; one on March 8, 1875; and one on August 11, 1879.

50. Franklin Notebook, Constitution and By-Laws.

51. *Green Bay Advocate,* March 23, 1871.

52. Guardian Treasurer's Report, February 14, 1876, 28; ibid., October 14, 1878, 79.

53. *State Gazette*, October 12, 1887.

54. Washington Log Book, January 1, 1889 and January 1, 1890.

55. *Green Bay Advocate,* May 10, 1883; *Green Bay Advocate,* June 17, 1886.

56. *Green Bay Advocate,* September 19, 1878.

57. Green Bay Common Council, 1883-86, 90.

58. Carl Herrmann, oral history November 10, 1936, 223 Main Street, Green Bay, NPM.

59. Green Bay Common Council, 1875-79, 393, and 660; Green Bay Common Council, 1880-83, 328; *Green Bay Advocate,* September 7, 1882.

60. *Green Bay Advocate,* March 6, 1884.

61. *Green Bay Advocate,* April 24, 1884.

62. *Green Bay Advocate,* October 5, 1885.

63. Green Bay Common Council, 1859-70, 218 and 246. For 1864; Green Bay Ordinances, 1854-75, Ordinance No. 146, November 18, 1867. For 1867; Green Bay Common Council, 1870-75, 162. For 1872.

64. Green Bay Ordinances, 1854-75, Ordinance No. 146, November 18, 1867.

65. *Charter of the City of Green Bay: All General Ordinances Passed Since the Organization of the City; To April 1st, 1872* (F. Burkard, 1872), Ordinance 14, 103.

66. *Green Bay Advocate,* March 14, 1872.

67. *Green Bay Advocate,* February 19, 1880; Sanborn, 1879, 2. Map shows congested building at this street corner.

68. Green Bay Ordinances, 1875-92, Chapter VIII, October 4, 1875, 60.

69. Green Bay Common Council, 1870-75, 571.

70. Green Bay Ordinances, 1875-92, Chapter VIII, October 4, 1875, 60.

71. *Charter and Consolidated Ordinances of the City of Green Bay with Amendments Thereto* (Green Bay: Rummel and Bender, 1882), 40, 120, and 123, UWGB-ARC.

72. *Green Bay State Gazette,* May 8, 1875; *Green Bay Advocate,* March 18, 1875.

73. Green Bay Common Council, 1880-83, 188. Quote; Green Bay Common Council, 1883-86, 8 and 13; Guardian Record Book, August 6, 1883, 167; *Green Bay Advocate,* December 6, 1883.

74. Green Bay Common Council, 1880-83, 429; Green Bay Common Council, 1883-86, 2 and 30; *Green Bay Advocate,* November 8, 1883.

75. Green Bay Common Council, 1880-83, 239.

76. Green Bay Common Council, 1880-83, 321; Green Bay Common Council, 1883-86, 30.

CHAPTER 6
THE GREAT FIRE OF 1880

1. John Atkinson and the Phenix Insurance Company, Plaintiff and Respondent vs. The Goodrich Transportation Company, Defendant and Appellant, State of Wisconsin, *Supreme Court, Cases and Briefs: Volume 178; 1882,* Wisconsin State Law Library, KFW 2400 B7 CB v. 178, 11 and 18 (hereafter cited as Atkinson, 1882).

2. Atkinson, 1882, 20. Quote; John Atkinson and the Phenix Insurance Company, Plaintiff and Respondent vs. The Goodrich Transportation Company, Defendant and Appellant, *Reports of Cases Argued and Determined in the Supreme Court of the State of Wisconsin: Volume LX, February 19-September 23, 1884.* Frederic K. Conover, Official Reporter (Chicago: Callaghan and Company, 1885), 141 (hereafter cited as Supreme Court, 1884). Wind estimate.

3. Atkinson, 1882, 5, 9, 11, 14, and 17.

4. Atkinson, 1882, 6.

5. Atkinson, 1882, 10. Quote; ibid., 7.

6. Atkinson, 1882, 17.

7. Atkinson, 1882, 12.

8. Atkinson, 1882, 21.

9. Supreme Court, 1884, 141.

10. Atkinson, 1882, 18.

11. Atkinson, 1882, 23

12. Atkinson, 1882, map. Citation refers to a map attached to the front page of the court case.

13. Atkinson, 1882, 28 and 31.
14. Atkinson, 1882, 24. Quote; ibid., map. Distance to mill.
15. Atkinson, 1882, map.
16. Atkinson, 1882, 25 and 28; Supreme Court, 1884, 141
17. Atkinson, 1882, 29.
18. Atkinson, 1882, 25-7.
19. *Green Bay Advocate*, September 23, 1880.
20. *Green Bay Globe*, September 22, 1880.
21. Atkinson, 1882, 34.
22. Atkinson, 1882, map. Distance between cistern and river; ibid., 75. Capacity of cistern calculated from dimensions.
23. Atkinson, 1882, 75.
24. Atkinson, 1882, 10 and 28.
25. Atkinson, 1882, 75.
26. Atkinson, 1882, 75.
27. Atkinson, 1882, 75.
28. Atkinson, 1882, 75.
29. Atkinson, 1882, 35.
30. *Green Bay Globe*, September 22, 1880.
31. Atkinson, 1882, 71.
32. *Green Bay Globe*, September 22, 1880.
33. Atkinson, 1882, 27.
34. Atkinson, 1882, 31.
35. Atkinson, 1882, 33.
36. Atkinson, 1882, map; *Green Bay Globe*, September 22, 1880; *Green Bay Advocate*, September 23, 1880.
37. Atkinson, 1882, 32.
38. Atkinson, 1882, 34. Elapsed time; ibid., map. Shows eight houses lost.
39. Atkinson, 1882, map.
40. Atkinson, 1882, 34; *Green Bay Advocate*, September 23, 1880. Garon home saved; Atkinson, 1882, map. Eight house on the same block burned.
41. Atkinson, 1882, 35.
42. Atkinson, 1882, map.
43. Atkinson, 1882, 36.
44. Atkinson, 1882, 37.
45. Atkinson, 1882, 35.
46. *Green Bay Advocate*, September 23, 1880.
47. *Green Bay Advocate*, September 23, 1880.
48. *Green Bay Advocate*, September 23, 1880.
49. *Green Bay Advocate*, September 23, 1880. Quote comparing wind to a tornado.
50. *Green Bay Globe*, September 22, 1880.
51. Atkinson, 1882, 36.
52. Atkinson, 1882, 33.
53. Atkinson, 1882, 39.
54. Atkinson, 1882, 36, 38, and 47.
55. Ian Griffith, Berner Schober Architects, e-mail message to author, February 22, 2015. Cedar-wood shingles (fairly flammable) were the most common roofing material in 1880, and hence most likely to have been on the Kitchen house.
56. Atkinson, 1882, 39.
57. Atkinson, 1882, 38
58. Atkinson, 1882, 44.
59. Atkinson, 1882, 44.
60. *State Gazette*, September 25, 1880.
61. *Green Bay Advocate*, September 23, 1880.
62. *State Gazette*, September 25, 1880. Quote; Atkinson, 1882, map. Block where six buildings burned.

63. *Green Bay Advocate*, September 23, 1880.

64. *State Gazette*, September 25, 1880.

65. Supreme Court, 1884, 141.

66. Supreme Court, 1884, 141; *State Gazette*, September 25, 1880; *Green Bay Globe*, September 22, 1880. Quote.

67. *Daily State Gazette*, September 22, 1880.

68. *Green Bay Advocate*, September 23, 1880.

69. *State Gazette*, September 25, 1880.

70. *Green Bay Advocate*, September 23, 1880; *Green Bay Globe*, September 22, 1880.

71. *Green Bay Globe*, September 22, 1880.

72. *Green Bay Advocate*, September 23, 1880.

73. *Green Bay Globe*, September 22, 1880; *Daily State Gazette*, September 22, 1880.

74. *Daily State Gazette*, September 22, 1880.

75. *Green Bay Globe*, September 22, 1880.

76. *Green Bay Advocate*, September 23, 1880; *Green Bay Press-Gazette*, September 18, 1925.

77. *Green Bay Advocate*, December 2, 1880; *Green Bay Globe*, September 29 and October 6, 1880.

78. *State Gazette*, September 25, 1880.

79. *Wisconsin Centennial Story of Disaster and Other Unfortunate Events: 1848-1948* (Wisconsin State Centennial Committee, 1948), accessed July 2, 2014, http://babel.hathitrust.org/cgi/pt?id=wu.89060458155;view=1up;seq=18; Daily State Gazette, April 29, 1875.

80. *Green Bay Advocate*, October 7, 1880.

81. *Green Bay Advocate*, September 23, 1880.

82. "Great Lakes Vessels Online Index," Bowling Green State University, accessed February 12, 2015, http://greatlakes.bgsu.edu/vessel/view/004015.

**CHAPTER 7
GREEN BAY AND FORT HOWARD
WATER WORKS COMPANY**

1. *Green Bay Advocate*, May 13, 1886.

2. Green Bay Common Council, 1883-86, 280.

3. Green Bay Common Council, 1883-86, 282. Committee formed; ibid., 303.

4. *Green Bay Advocate*, May 20, 1886; Fort Howard Common Council, 1886-94, 43-51.

5. Atkinson, 1882. Extensive description of failed water supply. See Chapter 6.

6. Green Bay Common Council, 1886-92, 2.

7. Fort Howard Common Council, 1886-94, 53.

8. Green Bay Ordinances, 1875-92, Water Works ordinance, July 19, 1886, 152-164. Company name.

9. Ibid.; Fort Howard Ordinances, 1886-94, Ordinance No. 115, August 18, 1886, 1-16.

10. *Green Bay Including Fort Howard Wisconsin* (New York: Sanborn Map, July 1887), BCL (hereafter cited as Sanborn, 1887). Insurance map showing hydrant locations in those areas with water mains.

11. *Green Bay Advocate*, July 29, 1886.

12. *Green Bay Advocate*, August 26, 1886.

13. *Green Bay Advocate*, September 9, 1886.

14. *Green Bay Advocate*, September 16, 1886.

15. *Green Bay Advocate*, October 7, 1886.

16. *Green Bay Advocate*, October 14, 1886.

17. *Green Bay Advocate*, June 30, 1887.

18. *Green Bay Advocate*, October 28, 1886.

19. *Green Bay Advocate*, December 2, 1886.

20. *Green Bay Advocate*, December 23 and December 30, 1886.

21. *Green Bay Advocate*, January 13, 1887.

22. *Green Bay Advocate*, February 3, 1887.

23. *Green Bay Advocate,* January 20, January 27, February 17, and February 24, 1887.

24. *Green Bay Advocate,* February 17 and March 3, 1887.

25. *Green Bay Advocate,* March 3, 1887.

26. *State Gazette,* May 25, 1887; *Green Bay Advocate,* June 9, 1887.

27. *State Gazette,* May 25, 1887.

28. *Green Bay Advocate,* June 30, 1887.

29. *Green Bay Advocate,* June 30, 1887.

30. *Green Bay Advocate,* July 28, 1887.

31. *Green Bay Advocate,* June 30, 1887.

32. *Green Bay Advocate,* July 28, 1887.

33. *State Gazette,* August 31, 1887.

34. Green Bay Ordinances, 1875-92, Water Works ordinance, July 19, 1886, 152-164; Fort Howard Ordinances, 1886-94, Ordinance No. 115, August 18, 1886, 1-16.

35. *State Gazette,* August 24 1887.

36. *State Gazette,* October 26, 1887.

37. *State Gazette,* December 14, 1887.

38. Fort Howard Common Council, 1886-94, 463.

39. *State Gazette,* June 1 1892.

40. Fort Howard Common Council, 1886-94, 537, 550, and 552.

41. *Green Bay Advocate,* September 1, 1887.

42. *State Gazette,* October 5, 1887.

43. *Green Bay Advocate,* August 2, 1888.

44. *Green Bay Advocate,* September 27, 1888.

45. *Green Bay Advocate,* August 16, 1888.

46. *Green Bay Advocate,* October 6, 1887.

47. *Green Bay Advocate,* February 9, 1888.

48. *Green Bay Advocate,* May 19, 1887.

49. *Green Bay Advocate,* September 20, 1888.

50. Green Bay Common Council, 1886-92 174. Mayor Neville proposed selling one of two steamer; ibid., 450. Discussion to send to a steamer to mill; Fort Howard Common Council, 1886-94, 118 and 128. Discussion on steamer, but no action.

51. *Green Bay Advocate,* March 11, 1886.

52. Green Bay Common Council, 1886-92, 122.

53. Green Bay Common Council, 1886-92, 181.

54. Fort Howard Common Council, 1886-94 112.

55. Green Bay Common Council, 1886-92, 113, 135, and 142; *Daily State Gazette,* October 27, 1887. Two hose carts.

56. Franklin Notebook, December 15, 1887; *Daily State Gazette,* October 27 1887.

57. Fort Howard Common Council, 1886-94, 124 and 179.

58. *State Gazette,* November 23 1892; *Green Bay Advocate,* November 24, 1892.

59. Sandborn map, 1887.

60. City of Green Bay-Water Department, *Plant Unit Ledger,* "Hydrant Rental Record, Account No. 348, Sheet numbers 1-4, Name of Units: 5" Hydrants," Green Bay Water Utility archives.

61. Green Bay Common Council, 1892-97, 272.

62. *State Gazette,* May 27, 1891. Second well; Insurance Maps Green Bay Brown Co. Wisconsin (New York: Sanborn Map, 1907), key sheet, BCL. Seven wells; The Insurance Year Book, 1916-1917, Fire and Marine, Fortieth-Fourth Annual Issue, Carefully Corrected to June 20, 1916 (New York: Spectator, 1917), C-409, http://babel.hathitrust.org/cgi/pt?id=wu.89095806717;view=1up;seq=5. Eleven wells.

63. National Board of Fire Underwriters, Committee on Fire Prevention, untitled map of Green Bay showing water supply system, fire stations, and recommended fire apparatus, December, 1915, Green Bay Water Utility archives.

64. *Green Bay Semi-Weekly Gazette,* June 23, 1906.

65. *Green Bay Semi-Weekly Gazette,* June 27, 1906. Site location; ibid., April 24, 1907. Opening.

CHAPTER 8
BIRTH OF THE FULL-TIME, PAID GREEN BAY FIRE DEPARTMENT

1. J. H. Beers, *Commemorative Biographical Record of the West Shore of Green Bay, Wisconsin: Including the Counties of Brown, Oconto, Marinette and Florence, Containing Biographical Sketches of Prominent and Representative Citizens and of Many of the Early Settled Families. Illustrated* (Chicago: J. H. Beers, 1896), 132.

2. *State Gazette*, July 6 1878.

3. Beers, *Commemorative Biographical*, 132. Civic activity; Green Bay Common Council, 1880-83, 169. Member of Fire Department Committee; *State Gazette*, October 6, 1883. Involvement with Business Men's Association; *Charter and Consolidated Ordinances of the City of Green Bay with Amendments Thereto* (Green Bay: Rummel and Bender, 1882), II, UWGB-ARC. Terms as alderman.

4. Guardian Secretary's Roll, May 29, 1863. Joins and listed repeatedly thereafter.

5. Guardian Record Book, inside back cover. Elected company secretary; *Directory of the Cities of Green Bay, Ft. Howard, De Pere and Nicolet 1884* (Milwaukee: A. G. Wright, 1884), 10, BCL. Listed as chief engineer; Green Bay Common Council, 1883-1886, 51 and 64. Britton was chief engineer for only one month. In his resignation letter, he cited the requirements of his business as not allowing adequate time for the fire department.

6. *State Gazette*, December 3, 1890. *Green Bay Advocate*, December 4, 1890. In depth descriptions of fire.

7. *State Gazette*, December 3, 1890; *Brown County Democrat*, December 4, 1890.

8. *State Gazette*, December 3, 1890; "The Inflation Calculator," accessed December 24, 2014, http://www.westegg.com/inflation/. For calculation of damage cost from 1890 as 2015 value.

9. *State Gazette*, December 3, 1890.

10. *State Gazette*, December 3, 1890.

11. *State Gazette*, December 3, 1890.

12. *State Gazette*, January 7, 1891.

13. *Green Bay Advocate*, December 4, 1890.

14. *Green Bay Advocate*, December 4, 1890; *State Gazette*, December 3, 1890.

15. *Green Bay Advocate*, December 4, 1890. Fort Howard steamer malfunction; Fort Howard Common Council, 1886-94, 355 and 358. Inquiries and repairs on steamer.

16. *Green Bay Advocate*, December 4, 1890. Britton at fire early.

17. Green Bay Ordinances, 1875-92, Water Works ordinance, July 19, 1886, 152-164.

18. *State Gazette*, December 17, 1890. Twelve streams used; *Green Bay Advocate*, December 4, 1890. First quote; *State Gazette*, December 3, 1890. Second quote.

19. *Green Bay Advocate*, December 4, 1890. Quote and Britton's estimate.

20. Green Bay Common Council, 1886-1892, 440.

21. Green Bay Common Council, 1886-1892, 441.

22. *State Gazette*, December 17, 1890.

23. *State Gazette*, February 24, 1892.

24. Green Bay Common Council, 1886-1892, 441; *Green Bay Advocate*, December 11, 1890.

25. *Green Bay Advocate*, December 11, 1890.

26. *State Gazette*, December 14, 1887.

27. *State Gazette*, December 10, 1890.

28. *State Gazette*, January 21, 1891. Extensive description of surprise water works test.

29. *State Gazette*, January 21, 1891.

30. Britton, Appellant, v. The Green Bay and Fort Howard Water Works Company, Respondent, 81 Wis. 48; 51 N. W. 84; (Supreme Court of Wisconsin, 1892).

31. *State Gazette*, September 3, 1890.

32. *Green Bay Advocate*, July 11, 1878; *State Gazette*, July 6, 1878.

33. Green Bay Common Council, 1886-1892, 442. Common Council request to fire company secretaries; *State Gazette*, December 17, 1890.

34. Green Bay Common Council, 1886-1892, 448.

35. *State Gazette*, December 17, 1890.

36. *State Gazette*, December 17, 1890.

37. Guardian Record Book, June 8, 1890. Only seven members total with Guardian No. 2; ibid., May 5. 1890, Four new members join Guardian No. 2.

38. Guardian Record Book, May 5. 1890.

39. City Directory, 1889-90, 34.

40. Green Bay Common Council, 1886-1892, 445; *State Gazette*, December 17, 1890.

41. Green Bay Common Council, 1886-1892, 442.

42. *State Gazette*, February 4, 1891.

43. *State Gazette*, February 18, 1891.

44. Green Bay Common Council, 1886-1892, 492. Salaries for two current engineers tallied as part of fire department reorganization; *The Insurance Year Book. 1891-1892 [Fire and Marine.] (Nineteenth Annual Issue.) Carefully Corrected to June 15, 1891.* (New York: Spectator, 1891), 245, http://babel.hathitrust.org/cgi/pt?id=umn.31951002481990p;view=1up;seq=11. Two full time firefighters listed for Green Bay; Carl Herrmann, oral history, November 10, 1936, 223 Main Street, Green Bay, NPM. Former chief stated Green Bay had two full-timers.

45. Green Bay Common Council, 1886-1892, 457.

46. Green Bay Common Council, 1886-1892, 459; *State Gazette*, April 1, 1891.

47. Washington 1885-92, March 4, 1891.

48. *State Gazette*, April 1, 1891.

49. Green Bay Common Council, 1886-1892, 459.

50. *State Gazette*, April 15, 1891.

51. Green Bay Common Council, 1886-1892, 470.

52. Green Bay Common Council, 1886-1892, 490.

53. Green Bay Common Council, 1886-1892, 490-2.

54. Green Bay Common Council, 1886-1892, 492.

55. *State Gazette*, May 27, 1891.

56. Green Bay Common Council, 1886-1892, 499.

57. Green Bay Common Council, 1886-1892, 519.

58. *State Gazette*, September 16, 1891; *Green Bay Advocate*, September 17, 1891; *State Gazette*, September 23, 1891.

59. *State Gazette*, September 23, 1891. Describes picture taken of Old Crocodile (this name used) surrounded by retired volunteer firefighters; *Green Bay Advocate*, September 24, 1891. Extensive description of final review activities.

60. *State Gazette*, September 23, 1891; *Green Bay Advocate*, September 24, 1891.

61. *State Gazette*, September 23, 1891. C.E. Crane erroneously recorded as O. E. Crane. Correct name found in numerous city directories and Federal Census for that time frame. No O. E. Crane found in Green Bay.

62. *State Gazette*, September 23, 1891. Two short quotes; *Green Bay Advocate*, September 24, 1891. Large quote.

63. *State Gazette*, September 16, 1891.

64. *Green Bay Advocate*, September 24, 1891.

65. Green Bay Common Council, 1886-1892, 535.

66. Green Bay Common Council, 1886-1892, 544. Kennedy appointed; *The Insurance Year Book, 1891-1892 [Fire and Marine] (Nineteenth Annual Issue), Carefully Corrected to June 15, 1891* (New York: The Spectator Company, 1891), 245, http://babel.hathitrust.org/cgi/pt?id=umn.31951002481990p;view=1up;seq=11. Herrmann listed as GBFD chief, but only for this edition.

67. *Daily State Gazette*, February 8, 1892; State *Gazette*, February 10, 1892; Green Bay Common Council, 1886-1892, 546.

68. Green Bay Common Council, 1892-97, 19, 25, 42, 49, 62, and 68.

69. *Directory of the Cities of Green Bay and Ft. Howard, 1893* (Chicago: R. S. Radcliffe, 1893), 6, BCL (hereafter cited as City Directory, 1893); *The Insurance Year Book, 1892-1893 [Fire and Marine] (Twentieth Annual Issue), Carefully Corrected to June 15, 1892* (New York: Spectator, 1892), 275, http://babel.hathitrust.org/cgi/pt?id=umn.31951002481991n;view=1up;seq=5. Eleven full-timers plus chief listed.

70. Green Bay (Wis.) Fire Department, "Log Books and Fire Record Books: 1899-1993", UWGB-ARC, Brown series 176, reel 1, 1899. Used daily roster entries to determine work schedule. Earliest found was from August 1899.

71. *State Gazette*, February 10, 1892; *State Gazette*, February 17, 1892.

72. *Daily State Gazette*, April 9, 1892.

73. *State Gazette*, May 11, 1892.

74. *Green Bay Advocate*, June 2, 1892.

75. Green Bay Common Council, 1892-1897, 66. Last of series of small payments to Franklin, Germania, Guardian, and Washington companies. Purpose for these payments not recorded.

76. *State Gazette*, March 9, 1892. Extensive description of new Engine House No. 2.

77. *State Gazette*, March 2, 1892.

78. Green Bay Common Council, 1892-97, 80; *Daily State Gazette,* March 4 and March 6, 1893. Direct references to bills for horses.

79. *State Gazette*, May 11, 1892.

80. *State Gazette*, January 28, 1891.

81. Green Bay Common Council, 1886-1892, 453 and 459; ibid., 490-492. Alarm system part of comprehensive re-organization of fire department.

82. *State Gazette*, January 20, 1892.

83. *State Gazette*, June 1, 1892. For box location.

84. City Directory, 1893, 6.

85. *State Gazette*, January 20, 1892.

86. Green Bay Common Council, 1886-1892, 545.

87. *State Gazette*, June 1, 1892.

88. *State Gazette*, June 1, 1892.

89. Green Bay Common Council, 1886-1892, 490-492.

90. Sanborn, 1887, 9. Location and size of tannery; *State Gazette*, June 1, 1892; *Green Bay Advocate*, June 2, 1892.

91. *Green Bay Advocate*, June 2, 1892.

92. *State Gazette*, June 1, 1892.

93. *State Gazette*, September 6, 1893; *Green Bay Advocate*, September 7, 1893.

CHAPTER 9
CREATION OF THE FORT HOWARD FIRE DEPARTMENT

1. Barton L. Parker, "The History and Location of Fort Howard", *Green Bay Historical Bulletin*, Brown County Historical Society, vol. 5, no. 4, October, November, December 1929.

2. Louise Phelps Kellogg, "The Story of Old Fort Howard: An Official Souvenir of the Old Fort Howard; Exhibit At The Wisconsin Tercentennial" (Green Bay: Tercentennial, July-August 1934).

3. Deborah B. Martin, *History of Brown County Wisconsin: Past and Present* (Chicago: S. J. Clarke, 1913), 1:297. Meeting.

4. French, *American Sketch Book,* 139.

5. Martin, *History of Brown County,* 300. Size; Fort Howard Common Council, 1856-65, 26. Number of votes.

6. Betsy Foley, *Green Bay: Gateway to the Great Waterway* (Woodland Hills, CA: Windsor Publications, 1983), 43.

7. Fort Howard Common Council, 1856-65, 8.

8. Fort Howard Common Council, 1856-65, 12.

9. *Daily State Gazette,* September 18, 1891.

10. Fort Howard Common Council, 1856-65, May 3 1858.

11. Fort Howard Common Council, 1856-65, June 18, 1858.

12. Fort Howard Common Council, 1856-65, June 18 and July 3, 1858.

13. Fort Howard Common Council, 1856-65, July 13, 1858.

14. Fort Howard Common Council, 1856-65, August 2, 1858.

15. Fort Howard Common Council, 1856-65, December 20, 1858. At Fisk warehouse; Brown County Tax Rolls, Fort Howard, 1860-1872, "Assessment Roll of the 1st Ward of the Borough of Fort Howard," UWGB-ARC, Brown series 4, box 32a, file folder "1863, Fort Howard First Ward."

16. Fort Howard Common Council, 1856-65, December 20, 1858.

17. Fort Howard Common Council, 1856-65, December 20, 1858.

18. J. B. Collins to Quartermaster General Thomas Jesup, October 7, 1849, NARA; Francis Lee to E. Backus, January 11, 1850, NARA.

19. "War Department Notes," Fort Howard Book 1, On Broadway, Inc., historical archive, Green Bay. Fort inactivate as of June 1852; Parker, 1929. Shaler as caretaker.

20. Fort Howard Common Council, 1856-65, November 24 and December 20, 1858.

21. Fort Howard Common Council, 1856-65, December 20, 1858.

22. Fort Howard Common Council, 1856-65, February 7, 1859. Bill paid; ibid., July 18, 1859.

23. Fort Howard Common Council, 1856-65, August 15, August 18, and September 3, 1859.

24. Fort Howard Common Council, 1856-65, August 3, 1859.

25. Fort Howard Common Council, 1856-65, November 12, 1860. Bill from Button and Blake; *The Encyclopedia of American Hand Fire Engines* (Handtub Junction, Fall 2001), 9 and 12. Button and Blake as manufacturer.

26. *Daily State* Gazette, September 18, 1891.

27. Fort Howard Common Council, 1856-65, May 23 1859.

28. Fort Howard Common Council, 1856-65, November 30, 1859 and January 4, 1860.

29. Fort Howard Common Council, 1856-65, June 6, 1860.

30. Fort Howard Common Council, 1856-65, August 4, 1860. Meades offers lot.

31. Brown County Clerk, "Tax and Land Records," UWGB-ARC, Brown series 11, box 2, vol. 6. Meades' lot number; C. M. Foote & W. S. Brown, Surveyors and Draughtsmen, "Plat Book of Brown County Wisconsin: Drawn from actual Surveys and the County Records" (Minneapolis: C.M. Foote, 1889), UWGB-ARC. Shows Meades' lot on South Pearl; Fort Howard Common Council, 1856-65, September 3, November 12, and December 3, 1860. Engine house moved.

32. Fort Howard Common Council, 1856-65, January 4, 1860, December 3, 1860, December 6, 1861, January 6, 1863 (quote), and December 11, 1863; Fort Howard Common Council, 1865-76, June 24, 1865.

33. Fort Howard Common Council, 1856-65, December 21, 1863.

34. *Green Bay Advocate,* December 24, 1868.

35. *Green Bay Advocate,* January, 21, 1869.

36. Fort Howard Common Council, 1865-76, October 2, 1867.

37. Fort Howard Common Council, 1865-76, January 6, 1869.

38. Fort Howard Common Council, 1865-76, February 3, 1869; A. Brauns, *Map of the Cities of Green Bay and Fort Howard* (Philadelphia: Wm. Brucher, January 1890). Used to determine modern street names. Borough records used pre-1895 annexation streets names.

39. Fort Howard Common Council, 1865-76, March 4, 1869.

40. Fort Howard Ordinances, 1868-94, Ordinance 63, November 11, 1869, 24.

41. Fort Howard Common Council, 1865-76, March 21, 1872; City Directory, 1874, 39; Conway, *Steam Fire Engines*, 236. Shipping record; *Green Bay Advocate*, November 2, 1871. Price paid.

42. *Green Bay Advocate,* March 21, 1872. Name.

43. Conway, *Steam Fire Engines,* 104.

44. Fort Howard Common Council, 1865-76, April 8, 1871.

45. Julie A. Hintz, Director, Waupaca Historical Society, emails to author, July 29 and August 12, 2014; June Johnson, Waupaca Historical Society, emails to author, August 12, August 13, and August 17, 2014.

46. *Green Bay Advocate,* June 1, December 14, 1871, and March 21, 1872.

47. Fort Howard Common Council, 1865-76, April 12, 1871.

48. Fort Howard Common Council, 1865-76, December 2, 1871.

49. *Green Bay Advocate,* December 14, 1871.

50. *Green Bay Advocate,* May 3, 1860. First recorded instance of Fort Howard Fire Department responding into Green Bay.

51. *Green Bay Advocate,* August 30, 1855. Fire in Fort Howard is the first recorded mutual aid response. Many other instances occurred. See Chapter 11.

52. City Directory, 1874, 38.

53. Conway, *Steam Fire Engines,* 103.

54. *Fort Howard Herald,* January 9, 1875. Quote; *State Gazette,* January 23, 1875.

55. *Fort Howard Monitor,* November 9, 1876 and March 14, 1878; *Green Bay Advocate,* March 7, 1878.

56. *Fort Howard Monitor,* September 14, 1876.

57. Fort Howard Common Council, 1865-76, April 15, 1872.

58. *Fort Howard Herald,* April 13, 1875.

59. *Fort Howard Herald,* March 29, 1877.

60. Fort Howard Common Council, 1865-76, October 4, 1875.

61. Fort Howard Common Council, 1865-76, April 15, 1872.

62. Green Bay Common Council, 1859-70, 588; Green Bay Common Council, 1870-75, 181, and 187.

63. Fort Howard Common Council, 1865-76, April 14, 1873.

64. Fort Howard Common Council, 1865-76; Fort Howard Common Council, 1876-85; Fort Howard Common Council, 1886-94. Numerous entries all three references to engineer of the steamer as paid position until 1887.

65. Fort Howard Common Council, 1865-76, June 17 and June 24, 1872.

66. *Daily State Gazette,* June 14, 1872.

67. Fort Howard Common Council, 1865-76, June 24, 1872.

68. Fort Howard Common Council, 1865-76, July 1 1872.

69. Fort Howard Common Council, 1865-76, October 7, 1872.

70. Fort Howard Common Council, 1865-76, August 12, 1872.

71. Fort Howard Common Council, 1865-76, August 12, 1872.

72. Fort Howard Common Council, 1865-76, February 17, 1873.

73. *State Gazette,* November 20, 1872; *Janesville Gazette,* November 9, 1872.

74. Fort Howard Common Council, 1865-76, November 7, 1872.

75. Fort Howard Common Council, 1865-76, November 7, 1872. Sheridan testimony; J. V. Suydam and A. Braun, "Map of the Cities of Green Bay and Fort Howard: From the Late Official Surveys and Map" (Green Bay: State Gazette, 1874), Wisconsin Historical Society, accessed November 26, 2014, http://content.wisconsinhistory.org/cdm/singleitem/collection/maps/id/12910/rec/2. Railroad on Pearl Street.

76. *Green Bay Advocate,* December 5, 1872.

77. Fort Howard Common Council, 1865-76, November 25, 1872.

78. Fort Howard Common Council, 1865-76, January 6, 1873; *Green Bay Advocate,* November 14, 1872.

79. Fort Howard Common Council, 1865-76, November 8, 1872.

80. Fort Howard Common Council, 1865-76, February 17, April 2, and April 8, 1873.

81. *State Gazette,* February 22, 1873.

82. *Green Bay Press-Gazette,* October 12, 1937.

CHAPTER 10
GRADUAL CHANGES THE FORT HOWARD FIRE DEPARTMENT

1. Fort Howard Common Council, 1865-76, November 15, 1875. First instance of payment for "fireing" the steamer; Fort Howard Common Council, 1865-76. Many payments for fireing; Fort Howard Common Council, 1876-85. Many payments for fireing; Fort Howard Common Council, 1886-94, January 5, 1894, 679. Couple of payments for fireing; Fort Howard Common Council, 1894-95, September 21, 1894, 29. Last payment found for fireing the steamer.

2. Fort Howard Common Council, 1876-85, 49.

3. Fort Howard Common Council, 1876-85, 70. Price; *Fort Howard Monitor,* October 26, 1876. Size.

4. Fort Howard Common Council, 1876-85, 83.

5. Fort Howard Common Council, 1876-85, 503.

6. Fort Howard Common Council, 1876-85, 521.

7. Fort Howard Common Council, 1876-85, 634.

8. Fort Howard Common Council, 1876-85, 638.

9. Fort Howard Common Council, 1876-85, 634-5. Horse team and teamster duties.

10. Fort Howard Common Council, 1886-94, 68.

11. *The Insurance Year Book. 1883-84. Carefully Corrected to June 20, 1883.* (New York: Spectator, 1883), 363, http://babel.hathitrust.org/cgi/pt?id=nyp.33433082318662;view=1up;seq=11. Population; Fort Howard Common Council, 1876-85, 483. Quote.

12. *The Insurance Year Book. 1882-1883. Carefully Corrected to June 15, 1882.* (New York: Spectator, 1882), 320, http://babel.hathitrust.org/cgi/pt?id=nyp.33433082318670;view=1up;seq=9.

13. Fort Howard Common Council, 1876-85, 483.

14. Fort Howard Common Council, 1876-85, 641.

15. Fort Howard Common Council, 1865-76, August 19 and November 16, 1874; Sanborn map, 1887, 13. Cistern labeled as being fed from the river.

16. *Fort Howard Herald,* June 22, 1875.

17. *Brown County Herald,* April 18, 1878.

18. *Fort Howard Review,* March 20 1877.

19. Fort Howard Common Council, 1876-85, 174 and 178.

20. Fort Howard Common Council, 1876-85, 174; *Fort Howard Review,* July 9, 1878.

21. Fort Howard Common Council, 1876-85, 319 and 505; Suydam, *Official Surveys and Map.* Used to determine modern street names. Borough records used pre-1895 annexation streets names.

22. Fort Howard Common Council, 1876-85, 634 and 638.

23. Fort Howard Common Council, 1886-94, 3, 74, 184, 283, 359, 465, and 572. Common council paid salaries; *The Insurance Year Book. 1889-90. (Seventeenth Annual Issue.) Carefully Corrected to June 15, 1889* (New York: Spectator, 1889), 203, http://babel.hathitrust.org/cgi/pt?id=nyp.33433082318407;view=1up;seq=6. Ten part-time firefighters listed. Same in subsequent editions of Spectator book; City Directory, 1889-90, 48.

24. Fort Howard Common Council, 1876-85, 96.

25. Fort Howard Common Council, 1886-94, 109.

26. Fort Howard Ordinances, 1868-94, Ordinance No. 115, August 18, 1886, 1-16.

27. Fort Howard Common Council, 1886-94, 112.

28. *State Gazette,* March 13, 1886. Eliminating steamers part of argument for a water works; *State Gazette,* May 25, 1887. Steamer fire engines not used at Cherry Street fire.

29. Fort Howard Common Council, 1886-94, 118.

30. Fort Howard Common Council, 1886-94, 128.

31. Fort Howard Common Council, 1886-94, 259 and 679; Fort Howard Common Council, 1894-95, 29. Both reference bills for fireing the steamer.

32. *State Gazette*, September 20, 1893. GBFD brought to Fort Howard for large mill fire along river.

33. Fort Howard Common Council, 1865-76, March 28 1872. Lot number 87 purchased; *Green Bay Including Fort Howard Wisconsin* (New York: Sanborn, 1894), 19, BCL. Location on Fourth Street; *Fort Howard Herald*, June 22, 1875. Triangle obtained; Fort Howard Common Council, 1876-85, 205. Bell obtained.

34. Wisconsin, Secretary of State, "Tax-Exempt Property Record: 1872-1874," Wisconsin Historical Society, series 229.

35. Fort Howard Common Council, 1886-94, 116 and 124.

36. Fort Howard Common Council, 1886-94, 143, 145, and 170.

37. Fort Howard Common Council, 1886-94, 179.

38. Fort Howard Common Council, 1886-94, 183, 453, and 471.

39. Fort Howard Common Council, 1886-94, 283.

40. Fort Howard Ordinances, 1868-94, Ordinance No. 115, August 18, 1886, 1-16.

41. Fort Howard Common Council, 1886-94, 471.

42. Fort Howard Common Council, 1886-94, 137.

43. *Daily State Gazette*, December 6, 1887.

44. Fort Howard Ordinances, 1868-94, Ordinance No. 115, August 18, 1886, 1-16.

45. Fort Howard Common Council, 1886-94, 463.

46. Fort Howard Common Council, 1886-94, 537 and 552.

47. Fort Howard Common Council, 1886-94, 463. Quote.

48. Fort Howard Common Council, 1886-94, 561; Fort Howard Common Council, 1894-95, 21.

49. *The Insurance Year Book. 1888-89. (Sixteenth Annual Issue.) Carefully Corrected to June 15, 1888.* (New York: Spectator, 1888), 184, https://books.google.com/books. 1887 statistics; *The Insurance Year Book, 1895-1896, [Fire and Marine] (Twenty-Third Annual Issue), Carefully Corrected to June 15, 1895* (New York: Spectator, 1895), 330, https://books.google.com/books. 1895 statistics.

50. Fort Howard Common Council, 1886-94, 678.

51. Fort Howard Common Council, 1886-94, 727.

52. Fort Howard Common Council, 1886-94, 681.

CHAPTER 11
MUTUAL AID LEADS TO A MERGER

1. "Historic Broadway: Parts 1 & 2, HB 3-A, HB 4-A," On Broadway, Inc., historical archives, Green Bay. Ferry in 1842; *Wisconsin Republican,* March, 11, 1845. States ferry operating by 1845.

2. "Fort Howard Book 1," On Broadway, Inc., historical archive, Green Bay.

3. French, *American Sketch Book,* 167; Foley, *Great Waterway,* 56.

4. French, *American Sketch Book,* 242.

5. *Green Bay Advocate,* March 28, 1895.

6. Green Bay Common Council, 1853-59, 495-7; *Green Bay Advocate,* February, 24, 1859.

7. Green Bay Common Council, 1870-75, 231; Fort Howard Common Council, 1865-76, March 28, 1872. Vote results reported.

8. Green Bay Common Council, 1870-75, 371 and 379; *Green Bay Advocate,* March 26, 1874.

9. *Green Bay Advocate,* February, 14, 1889.

10. *Green Bay Advocate,* April, 11, 1895.

11. *Green Bay Advocate,* August 30, 1855.

12. *Green Bay Advocate,* July 7, 1859.

13. *Green Bay Advocate,* August, 21 and August 28, 1856.

14. *Green Bay Advocate,* December, 11, 1856.

15. *Green Bay Advocate,* July 21, 1859.

16. Green Bay Common Council, 1853-59, 548.

17. *Green Bay Advocate,* May 3, 1860.

18. *Bay City Press,* April 5, 1862.

19. *Green Bay Advocate,* December, 26, 1861.

20. *Green Bay Advocate,* June 9, 1864.

21. Green Bay Ordinances, 1854-75, Ordinance No. 45, March 27, 1858, 80; ibid., Ordinance No. 48, April 24, 1858, 86.

22. *Green Bay Advocate,* July 7, 1859.

23. *Green Bay Advocate,* September 1 and September 8, 1859.

24. *Green Bay Advocate,* January 21 and January 28, 1858.

25. *Green Bay Advocate,* September 27, 1860.

26. *Green Bay Advocate,* July 29, 1869.

27. *Green Bay Advocate,* July 7, 1870.

28. *Green Bay Advocate,* March 28, 1872.

29. *Green Bay Advocate,* November 28, 1872.

30. *Green Bay Advocate,* January, 16, 1879.

31. Fort Howard Ordinances, 1868-86, Ordinance No. 94. March 4, 1879, 323.

32. *Green Bay Advocate,* February 10, 1876; *Brown County Herald,* June 20, 1878.

33. *Green Bay Advocate,* August, 1, 1872.

34. *Green Bay Advocate,* October 31, 1872.

35. Fort Howard Common Council, 1865-76, June 14, 1872; *Green Bay Advocate,* June 20, 1872.

36. City Directory, 1874, 38.

37. Fort Howard Common Council, 1865-76, April 8, 1871.

38. *Green Bay Advocate,* March 21, 1872.

39. *State Gazette,* November 20, 1872; *Janesville Gazette,* November 9, 1872.

40. Fort Howard Common Council, 1865-76, November, 25, 1872.

41. Fort Howard Common Council, 1865-76, January, 6, 1873; *State Gazette,* February 22, 1873.

42. Green Bay Common Council, 1859-70, 244 and 246.

43. Fort Howard Common Council, 1865-76, February 3 and October 15, 1869.

44. Fort Howard Common Council, 1865-76, January 31, 1870. 1870 payment; Green Bay Common Council, 1870-75, 212. 1872 payment.

45. *Green Bay Advocate,* May 28, 1874.

46. Fort Howard Common Council, 1865-76, May 18, 1874; *Green Bay Advocate,* June 4, 1874.

47. *Green Bay Advocate,* June 4, 1874.

48. Green Bay Common Council, 1870-75, 414.

49. *Green Bay Advocate,* May 28, 1874.

50. *Green Bay Advocate,* May 28, 1874.

51. Green Bay Common Council, 1870-75, 414.

52. *Green Bay Advocate,* May 28, 1874.

53. *Green Bay Advocate,* June 4, 1874.

54. *Green Bay Advocate,* June 18 and June 25, 1874.

55. *Green Bay Advocate,* June 18, 1874.

56. *Green Bay Advocate,* June 18, 1874.

57. *Green Bay Advocate,* June 25, 1874.

58. *Green Bay Advocate,* August, 20, 1874.

59. *Fort Howard Herald,* August, 25, 1874.

60. *Green Bay Advocate,* September, 3, 1874.

61. *Fort Howard Herald,* January, 19, 1875. Quote; *Green Bay Advocate,* January, 21, 1875.

62. *State Gazette,* January, 23, 1875.

63. *Green Bay Advocate,* January, 21, 1875.

64. *Fort Howard Herald,* January, 26, 1875. Agreement details; Fort Howard Common Council, 1865-76, February, 1, 1875.

65. *Fort Howard Herald,* April 20, 1875.

66. *Green Bay Advocate,* April, 1 and May 13, 1875.

67. *Fort Howard Herald,* July 27 and August 25, 1875.

68. *Daily State Gazette,* August 21, 1875.

69. Green Bay Common Council, 1870-75, 635.

70. *Green Bay Advocate,* March 29, 1877.

71. *Green Bay Advocate,* October, 19, 1876.

72. *Green Bay Advocate,* October, 19, 1876.

73. Green Bay Common Council, 1875-79, 258 and 269; Fort Howard Common Council, 1876-85, 83 and 86.

74. *Green Bay Advocate,* March, 8, 1877.

75. *Green Bay Advocate,* March, 8, 1877.

76. *Green Bay Advocate,* March, 8 and March 15, 1877.

77. *Green Bay Advocate,* January 28, 1886.

78. *Green Bay Advocate,* March 15, 1877. GBFD report; Fort Howard Common Council, 1876-1885, 585. Fort Howard Fire Department report.

79. *Green Bay Advocate,* July 6, 1876. Parade; *Green Bay Advocate,* June 7, 1877; *Green Bay Advocate,* June 17, 1886.

80. *Green Bay Advocate,* September, 24, 1891; *Daily State Gazette,* September, 18, 1891.

81. *Fort Howard Monitor,* January 4, 1877.

82. *Green Bay Advocate,* May 15, 1879.

83. *Fort Howard Journal,* August 1, 1879; *Green Bay Advocate,* August 7, 1879.

84. *Stevens Point Journal,* July 26, 1879.

85. *Green Bay Advocate,* August 30, 1883.

86. Green Bay Common Council, 1870-75, 439, 568, 573, 617, 629, 648, 658, and 690; Green Bay Common Council, 1875-79, 29, 36, 96, 101, 105, 159, and 367. All refer to payments to Kittner by Green Bay.

87. *Green Bay Advocate,* March, 29, 1877.

88. Green Bay Common Council, 1875-79, 367.

89. Fort Howard Common Council, 1876-85, 122.

90. Fort Howard Common Council, 1876-85, 186, 188, 214, and 283.

91. Atkinson, 1882, 34.

92. *State Gazette,* December 17, 1890. Hired as GBFD Engineer of the Steamer No. 1; Green Bay Common Council, 1886-92, 546. Hired as full-time GBFD firefighter.

93. *State Gazette,* June 29, 1892.

94. *Directory of the Cities of Green Bay, Ft. Howard, De Pere and Nicolet 1884* (A. G. Wright, Milwaukee), BCL. Address; *Green Bay Including Fort Howard* (New York: Sanborn Map, November 1883), BCL, 13. Insurance map.

95. *State Gazette,* September, 8, 1883.

96. *State Gazette,* September, 8, 1883.

97. Green Bay Common Council, 1883-86, 19; *Green Bay Advocate,* September, 27, 1883.

98. Green Bay Common Council, 1883-86, 43-47; Fort Howard Common Council, 1876-85, 547-8.

99. Green Bay Common Council, 1883-86, 46.

100. Green Bay Common Council, 1883-86, 54.

101. Green Bay Ordinances, 1875-92, Water Works ordinance, July 19, 1886, 152-164; Fort Howard Ordinances, 1868-94, Ordinance No. 115, August 18, 1886.

102. Green Bay Common Council, 1892-97, 263.

103. *Green Bay Gazette,* April, 3, 1895; *Green Bay Weekly Gazette,* April 10, 1895.

104. Green Bay Common Council, 1892-97, 278.

105. Fort Howard Ordinances, 1886-94, March 1, 1895, 246; Green Bay Ordinances, 1893-1903, March 9, 1895, 58.

106. Fort Howard Common Council, 1886-94, 525 and 531.

107. Green Bay Common Council, 1892-97, 281.

108. *Green Bay Advocate*, May 9, 1895; *Green Bay Weekly Gazette*, May 15, 1895.

109. *Green Bay Weekly Gazette*, May 29, 1895.

110. *Green Bay Weekly Gazette*, May 15, 1895.

111. *Green Bay Gazette*, May 4, 1895; *Green Bay Advocate*, May 9, 1895.

112. *Green Bay Advocate*, May 9, 1895.

113. *Wright's Directory of Green Bay for 1896-7* (Milwaukee: A. G. Wright, 1896), 34, BCL (hereafter cited as City Directory, 1896-97). First city directory to list full-time firefighters on the west side

114. *Green Bay Gazette*, May 4, 1895; *Green Bay Advocate*, May 9, 1895.

115. Fort Howard Common Council, 1886-94, 678; City Directory, 1893, 30; *Wright's Directory of Green Bay Fort Howard for 1894-1895* (Milwaukee: A. G. Wright, 1894), 49, BCL.

116. Fort Howard Common Council, 1886-94, 660. Hired; Green Bay Common Council, 1892-97, 276. Green Bay Common Council paid last month salary as Fort Howard teamster.

117. Green Bay Common Council, 1892-97, 276 and 289.

118. City Directory, 1896-97, 34.

119. *Green Bay Gazette*, May 4, 1895.

120. *Green Bay Gazette*, May 4, 1895.

121. Green Bay Common Council, 1892-97, 289. Hose purchased; *Green Bay Advocate*, May 25, 1895; *Green Bay Weekly Gazette*, May 29, 1895.

122. City Directory, 1896-97, 35.

123. Green Bay Common Council, 1892-97, 296 and 591.

124. *Green Bay Weekly Gazette*, July 10, 1895.

125. Green Bay Common Council, 1892-97, 570.

126. Green Bay Common Council, 1892-97, 302. Quote. Ibid., 329.

127. Fort Howard Ordinances, 1886-94, March 1, 1895, 250; Green Bay Ordinances, 1893-1903, March 9, 1895, 60.

Image Credits

For book sources, see bibliography for full citation.

Newspaper images courtesy of the Brown County Library and the Wisconsin Historical Society

Use of Sanborn maps of Green Bay, Wisconsin for 1879, 1883, 1887. 1894, 1900, 1907, 1936, and 1956. Many have updates, but these are the original publication years. Reprinted/used with permission from The Sanborn Library, LLC.

Alpena County Library: 117

Author: 8, 23, 74, 75, 77, 97, 161, 183, 282, 288 (upper), 289-291, 294, 295, 298 (upper), 299, 300 (upper left), 305, 306 (upper), 307 (upper), 308 (lower), 309 (upper), 310 (upper), 311 (upper), 312 (upper), 313 (lower), 314 (right)

Beers, *Commemorative Biographical Record*: 169

Brown County Library: 34, 35, 43, 51, 102, 104, 105, 108, 126, 143, 147, 148, 159 (upper), 180, 212, 213, 238, 239, 246, 256, 308 (upper)

City Directory, 1872: 40

City Directory, 1874: 118

City Directory, 1884: 167

City Directory, 1889-90: 170

City Directory, 1893: 197

City Directory, 1896-7: 279

City of Green Bay: 5, 7, 47, 93, 94, 99, 107, 109, 144, 153, 159 (lower), 164 (right), 168, 177, 194, 199, 204, 234, 235, 243, 249, 264, 267, 273

Conway, *Those Magnificent Old Steam Fire Engines*: 86

De Pere Historical Society: 131

Dick and Don Newman (personal collection): 173

Dunshee, *Enjine!-Enjine! A Story of Fire Protection*: 13

Firemen's Association of New York Museum of Firefighting, Hudson: 211, 219

Green Bay Metro Fire Department archives: 71, 101, 106, 193, 195, 281, 306 (lower), 310 (lower), 311 (lower), 313 (upper), 314 (left)

Green Bay Press-Gazette: 119

Heritage Hill State Historical Park: 288 (lower)

Mike Hronek: 26, 58, 81, 90, 113, 140, 166, 202, 228, 250, 286, 304

Neville Public Museum of Brown County: 39, 42, 45, 46, 56, 65, 68, 72, 79, 82, 85, 96, 98, 110, 111, 132, 135-137, 154, 155, 158, 160, 162, 163, 165, 187, 196, 198, 201, 222, 257, 293, 296, 297, 298 (both lower), 300 (upper right and lower), 301-303, 307 (lower), 309 (middle and lower)

Page, *Illustrated Historical Atlas of Wisconsin*: 125, 231, 270

University of Wisconsin-Green Bay, Area Research Center (see endnotes for citation explanation): Fort Howard Common Council, 1856-65: 207; Green Bay Common Council, 1853-59: 27, 30, 38; Green Bay Ordinances, 1854-75: 41, 49, 52

United States Department of the Interior, Bureau of Land Management, Eastern States, www.glorecords.blm.gov: 2

William Gleason, GBFD Chief (personal collection): 261

Wisconsin Historical Society: WHS-111947 (map), 89 and 226; WHS-31586 (church), 120

Wisconsin State Law Library: 114, 123, 130

Index

(Page numbers with "i" refer to images.)

Ahnapee (Algoma), 22, 24, 94, 95i, 96i, 97i

Aldermen: Fort Howard, 224, 225, 244, 258, 261, 274; Green Bay, 49, 65, 64, 179, 200

Amoskeag, 68, 68i, 69, 71i

Annexation of Fort Howard, 252, 275, 275i

Annual inspection, review, parade, 40, 46, 50; 1891 final, 186i, 187-189, 187i; fine for missing, 102; firefighters in uniform, 79i, 85i, 101i, 160i; Fort Howard included, 257, 270; Water Works showcase, 156

Arrest powers, firefighter's, 41, 44, 60, 64-66

Artesian wells, 144, 144i, 146, 147, 246

Assistant Chief Engineer, 41, 54i, 64, 162i, 185; part time, 96-97; Fort Howard, 224, 258

Astor Planing Mill, 118, 118i, 119i, 122, 130; map 123i

Baily, T. F., 210

Beattie, James A., 225

Bell, fire alarm, 48, 76, 77, 77i; alerting water works, 152, 245; Astor sold, 93; donated, 76, 77i; Fort Howard, 214, 242, 243i; mutual aid, 257, 258, 266

Bell tower, 71, 76, 77; Fort Howard, 214, 242; images, 43i, 72i, 104i, 108i, 222i, 281i

Bertles, John, 110i, 121

Blesch Brewery, 36, 218, 253

Bridges: improving mutual aid, 59, 251, 255; water source, 214, 228i, 237, 239

Britton, D. W., 167, 168, 169i, 177, 178, 187i, 188, 200, 201; 1890 fire, 169-173, 171i; cooperage, 167, 167i, 168i, 170i, 178, 200, 201; first hydrant in Green Bay, 146; Water Works, 174-176

Brown County Fair Firemen's Competition (1860), 45, 46i, 173i, 256

Brown County Sheriff, papers destroyed (1841), 12

Buckets, bucket brigade, 9-13, 10i, 18, 62, 206; required, 9, 10, 207

Buildings, fire proof, 53, 61, 62

Button and Blake, hand pumpers, Guardian No. 2, 38, 42i; Fort Howard, 210, 211, 211i, 217

Button and Son, Fort Howard steamer, 216, 216i, 219, 219i,

Chief Engineer, 41, 50i, 54i, 64, 80, 110i, 162i; Fort Howard, 212, 236, 269; mutual aid, 255, 257, 258, 261, 263, 265, 268; nomination process, 97, 224; part-time, 96

Chief Warden, 50, 50i

Christensen, Arthur, Sr., 315

Cisterns, see water supply

Clapp and Jones, 78, 79i

Coal for steamers, 65i, 66, 217, 221, 229; wagon, 87, 88

Common Council, Fort Howard, 171; 1887 false alarm controversy, 244, 245; cisterns, other water sources, 214, 215, 237; construction and fire limits, 215-216; early fire regulations and authority, 206-208, 207i; engine house, 209, 211-214, 227; fire engines purchased, 208, 210, 216, 218; fire-the-steamer bills, 230; fire warden, 207; horse team bounty/contracts, 220, 221, 229, 232, 233, 241, 242, 247; hire engineer of the steamer, 222-223; Investigates engine house fire (1872), 225-227; mutual aid, 219, 261, 262; purchases, 242; reorganization of fire department (1872), 223-225; staffing issues, 158, 236, 240; steamer out-of-service, 241; water works, 142, 144, 246

Common Council, Green Bay: acknowledges volunteer companies, 32, 38i, 40, 44, 45, 72; approve nominations, 97; assigns fire wardens, 51; Britton fire, 172i, 173, 178, 179; cisterns and docks, 48, 73; disband Astor and Franklin, 91i, 92; disband volunteers, 190, 190i; dismiss engineer of the steamer, 99, 158; engine houses, 40, 41i, 46, 47; establishes full-timers, 183-185, 190; financial support, 99, 100; Fire Department Committee, 49, 152, 179, 183, 276, 277; fire limits ordinance, 52, 52i; fire warden ordinance, 49i, 50; Gamewell alarm, 196, 198; hand pumpers purchased, 24, 38, 67, 68, 77, 78; heating stations, 36, 47; horse team bounty, 48, 75, 76, 86; mutual aid, 254, 256, 260, 261, 268, 274; organize fire department, 24, 41, 41i; rooms at Station No. 1, 110; sell hand pumpers, 93, 94; water works, 142, 143, 162.

Cook's Hotel, 105i, 108

Crane, C. E., 188

Delwiche, August, 84

Docks, fire department: destroyed 1880 fire, 130; no longer needed, 158; water supply, 48, 80, 81, 81i, 105, 106, 113i

Dues and fines, volunteer firefighters, 100-103

Eastmen, F. E., 52

Eldred Lumber and Manufacturing Company, 270i, 272, 273i

Elmore, James, 175, 177i, 179, 183, 189, 200

Engineer of the Steamer, 110i, 173i; duties, 77, 258; eliminated, 158, 241; Fort Howard, 223, 224, 230, 240; hired Green Bay, 70, 78, 181

Ferry/boat, carrying firefighters across Fox River, 36, 253, 254, 256

Fines, on citizens: fire limits violations, 53, 216; not having fire buckets, 10; refusing fire warden inspection or direction, 17, 18; refusing to help on hand pumper, 41, 42, 60, 64;

Fire, mutual aid examples, 253-254; September, 1883 (Eldred Mill), 270i, 272-274, 273i

Fires, significant (chronologically): 1840, 12; December 24, 1841, 12; March 1850, 21; November 1, 1853, 28; November 28, 1854, 30; December 1854, 31; November 28, 1854, 30; November 11, 1863, 60-63; August 1866, 64, 66; September 20, 1880, 116-134; May 1887, 149, 150i; November 27, 1891, 169-173; May 24, 1891, 198-200

Fire alarm systems: bell system by wards, 77, 257, 258; Cook's Hotel, 105i, 108; municipal, see Gamewell

Fire company, Alert Fire Company, 24, 25, 25i, 96i

Fire company, Astor Fire Company No. 1, 76, 91, 97i; disbanded, 91i, 92; engine house, 82i, 93, 93i; formed, 72; hand pumper, 75i, 93, 96i

Fire company, Fire Company No. 1, 19i, 20-22, 20i

Fire company, first (unnamed), 14, 14i, 18, 19

Fire company, Fort Howard engine company, 46, 46i; steamer at 1880 fire, 131, 131i, 134

Fire company, Franklin Engine Company No. 3 (1860), 35i, 45, 46, 47, 47i; disbanded, 91i, 92; new hand pumper, 78, 85i

Fire company, Franklin Hose Company No. 3 (1887), 158i, 159, 159i, 160i; disbanded, 184

Fire company, Germania Fire Company No. 1, 35i, 36, 49, 110i; creation, 32, 32i, 33; nickname, 32, 69; resign, 183; steamer, 68i, 69.

Fire company, Guardian Fire Engine Company No. 2, 56i, 171i; banner, 39, 39i, 42i; Britton fire (1890), 171, 173i, 180; creation, 38, 38i; hand pumpers, 38, 42i, 256; hose cart, 95; in warehouse, 40, 41i; nickname, 38, 39i, 78, 79i; North Adams station, 35i, 43i 46, 47; parties, 49, 100i; steamer, 78, 78i, 79i, 271; purchase horses, 83

Fire company, Green Bay Sack and Protection Company, 44-46

Fire company, Howard Fire Co. No. 1, 208, 209

Fire company, Live Oak Company, 224, 225

Fire company, Resolute Fire Company, 160, 242-244, 243i, 267i

Fire company, Washington Hook and Ladder Company No. 1, 49, 76, 93, 101i, 256; Cherry St. house, 35i, 45i, 47; created, 40; North Adams house, 46; rented warehouse, 41i, 40, 46; South Washington station, 106i, 109; wagons, 40, 40i, 95, 101i

Fire Department Committee, see Common Council, Green Bay

Fire engine, Enterprise, Steamer No. 1, 68, 69, 71i, 73; at 1880 fire, 120, 124

Fire engine, hand pumper: brought across Fox, 13; Button and Blake, Fort Howard, 210-211, 211i; Button and Blake, Guardian No. 2, 38, 42i; description and operation, 8i, 13, 72; Franklin No. 3 (1870), 79, 85i; from US Army to Fort Howard, 208-210; from US Army to Green Bay, 13-16, 18; Old Croc, 8, 14, 17, 19, 38, 94, 98i, 187i, 188; manpower limitation, 15, 59, 60, 62-64; Smith, 22i, 24, 24i, 32, 33, 72, 75i, 96i, 294i; sold by Green Bay, 93, 94, 217; women operating, 63

Fire engine, steamer: bought by Fort Howard, 216, 217, 216i, 219i, 219; bought by Green Bay, 68, 69, 77, 78, 78i, 79i; description, 65i, 66, 68i, 71, 73, 74; fireing the steamer bounty, 230, 241, 271, 272; obsolete, 159, 241, 243; operation, 70, 71i, 86, 86i, 87

Fire wardens, 14, 17, 18, 60; ordinance, 49i, 50; in Fort Howard, 207, 209

Firemen's Hall, 50, 62, 100i

Fire station, Cherry Street, 35i, 45i, 47, 99i, 109, 112

Fire station, current No. 1, 71, 82i, 93i

Fire station, current No. 3, 279, 281i

Fire station, former, current views, 304-313

Fire station, Fourth Street (Resolute), 242, 243i, 267

Fire station, Main Street (Franklin): first (1860), 35i, 47, 47i; second (1887), 159, 159i, 160, 282i, 295i

Fire station, Main Street (No. 2), 184, 186, 193-195, 193i, 194i

Fire station, No. 4, 280

Fire station, North Adams, 35i, 43i, 109, 109i, 111i; built, 46, 47; sold, 184, 186, 187

Fire station, Pearl Street: burned, 225-227; first, 213, 213i; second 222i, 227, 235i, 239i, 246; taken over after merger, 264i, 276, 277i

Fire station, South Washington (1852), 33, 35i, 48, 71

Fire station, South Washington (1868), 71, 72i, 77, 181, 187i, 195, 260, 264i; keystone, 300i

Fire station, Walnut and Adams, 16, 17

Fire station, Washington and Adams, 71, 72, 82i, 93i

Fire station locations (maps), 26, 58, 90, 140, 166, 202, 228, 250, 304

First Presbyterian Church, 120i, 123, 126i

Fisk and Company warehouse, 209

Follet, Burley, 33

Fort Howard, US Army garrison: Old Croc, 8i, 12, 15, 98i; engine to Fort Howard, 21, 208-210

Fort Howard Fire Department, 208, 223-225, 235, 236; false alarm controversy, 244, 245; merger, 276-279

Fox River: crossed for mutual aid, 36, 59, 253-255; crossed to get Old Croc, 13; fire department docks, 80, 80i, 105; water mains across, 144, 146-149, 162; water source, 72, 80, 142, 214, 215, 234, 242;

Fuerst, Ed, 110i

Full-time, paid firefighters: engineer of the steamer, 70, 78, 158, 181, 241; in Fort Howard, 276i, 277, 277i; in Green Bay, 184, 185, 190, 191, 191i; working hours, 70, 191

Fundraiser, balls, parties, dances, 36, 49, 50, 100i, 103, 270, 271

Gamewell Fire Alarm Telegraph Company, 196, 198; system, 183i, 196i, 197i, 198i, 278i, 279i; west side, 278, 298i, 300i-303i

Goodrich Transportation Corporation, 139

Gray, A. L., 189, 208, 244, 269

Great Fire of 1880: effects, 133; firefighting efforts, 120, 121, 124-127, 131; maps, 114i, 123i, 129i, 130i; northern section, 128-130; origin and cause, 116-119; surviving buildings, 286-291; water supply problems, 120, 121

Green Bay, Borough: charter and fire department, 10; municipal records lost 28, 30i.

Green Bay Advocate destroyed, 28, 29

Green Bay and Fort Howard Water Works Company, see Water Works

Green Bay Business Men's Association: push for water works, 142, 143; response to Britton fire, 173-175

Index 347

Green Bay Fire Department: name use, 50; 1890 annual report, 170; organizing ordinance, 41, 41i

Green Bay Fire Department, full-time: Engine Co. No. 1 lineage, 32; established, 167, 185; merger with Fort Howard, 276-278

Green Bay Fire Department, volunteer membership: companies, 20, 24, 32, 39, 173i, 180; entire department, 40, 97-99, 181, 183

Green Bay Metro Fire Department, 272, 283; Station 1, 82i, 93i; Engine No. 2 lineage, 38i

Green Bay Police Department, 188

Green Bay Water Utility, 74i, 145, 245

Hagen, Frank, 84, 85, 274

Hannis, G. W., Company, 79i, 95

Hansen, Hans Mark, 191, 191i, 192i, 315

Harder, Homer, 128

Heritage Hill State Historical Park, 159, 160i, 161i, 294i, 295i, 299i

Harrington, Water Works Superintendent, 176

Herrmann, Carl, Chief, 175, 176, 179, 181-183, 187, 190

Hose, 48; drying tower, 71, 72i, 95, 108i; purchases, 36 48, 69, 75, 208, 209, 216, 278; supply hose, 66, 68i, 71i, 80, 239

Hose carts, 42i, 45, 75, 79i, 97i, 161i, 191, 281i, 294i; hauling, 178, 220, 221, 231, 232; hose cart only companies, 158i, 159, 160, 242, 243; purchases, 45, 69, 78, 95, 211; damaged, 273

Horse teams, 48, 86, 195i, 256i, 257i; stalls, 194, 276; contract in Fort Howard, 233, 240; Guardian purchase own, 83; purchased in Green Bay, 185, 195, 196; purchase in Green Bay rejected, 75, 76, 86; teamster, 234, 240; injured, 84, 273, 274

Horse teams, bounty program: cost, 267; Fort Howard, 84, 220, 221, 230-232, 240; Green Bay, 75, 83, 84, 87, 158; mutual aid agreement, 266; problems, 178, 260, 261, 265

Howe, T. O., 253

Hydrants, 146, 153i; cost, 146; first used, 149, 150i; number of, 144, 162; type 144, 154i, 155i; use without engine, 157, 159, 160

Ingalls, J., 54

Insurance premiums, payout to fire departments, 100, 182

Jackson Square, 123i, 124, 125, 127, 136i

Joppe, August, 315

Kennedy, Dan, 315

Kennedy, William, 190, 199, 276, 278

Kimball, A. Weston, 262

Kitchen, Charles (home), 128-131, 130i

Kittner, Ed, 110i

Kittner, John, 110i, 181, 190, 271, 272; 1880 fire, 120, 121

Klaus, Anton, 69

Klaus Hall, 50, 62, 100i

Klaus, Phillip, warehouse, 40

Lathrop, F. A., 32

Leicht, Theodore, 276i, 277

Lenz, Franz, 292i

Lindley, Samuel, 257

Loewert, Henry, 110i

Lucas, James, 70

Ludlum, Harry, 13i, 16, 94

Lutheran church, 130

Lumber yards, banned in fire limits area, 108

Matthews, Fred, 315

Mayors: Elmore, James (Green Bay) 175, 177i, 179, 183, 189, 201; Fort Howard, 223, 266; Green Bay, 31, 37, 64; mutual aid, 256, 258, 261, 263, 265

Methodist Episcopal Church, 125

Miller, Mike, 315

Mohr, Louis, 122

Morrow, Elisha, 21, 124, 126i

Municipal water system, see Water Works

Murphy, Frank, 200

Mutual aid, 59, 259; 1874 controversy, 261-265; 1880 fire, 119, 131, 134; first instance (with US military), 13; examples, 253-255, 258, 266, 268, 269, 272-274; formal agreement, 255, 256, 265, 266, 268; frequency, 269; horse bounty payment, 260; no engine in Fort Howard, 218, 259; steamers out of service, 271; system, 257, 258

Navarino, village meeting (1836), 9

Neville, A. C., 142, 154i

Neville Public Museum, 16, 39, 39i, 42i, 46, 46i, 196i, 198i

Nooyen, Frank, 315

Oconto, The (ship), 116-119, 117i, 139

Old Croc, 8i, 16-19, 293i; 1891 Final Review, 187i, 188; borrowed from fort, 13-14; current status, 94, 98i; purchased, 15-16, 18; used by Franklin, 45, 79; used by Guardian, 38

Ordinances: annexation, 275, 278i, 279; buckets, 9, 10, 11, 11i; construction rules, 11, 50, 53, 215, 216; fire limits, Fort Howard, 215; fire limits, Green Bay, 51i, 52, 52i, 61, 102i, 104i, 106, 107; fire wardens, 17, 31, 50, 49i, 60, 209; full-time, paid, 184, 185; help on hand pumpers, 64; moving wooden buildings prohibited, 53; mutual aid, 255, 256, 258; organize fire companies, 20, 20i, 44; organize fire department, 41, 41i, 76, 98; water works, 144, 145i, 149, 151, 172, 174

Parades, other than annual inspection, 25, 38, 42i, 49, 256, 257, 270

Pfotenhauer, Charles, 110i, 162i

Phoenix newspaper destroyed, 12

Pine Street School, 131

Post office destroyed, 62

Rahr Brewery, 159i, 199, 199i

Reber, Henry, 133, 162i

Reservoirs, see Water Works

Richardson, George, 237, 266

Ritchie, William 83

Robinson, Charles, 67, 69

Salaries: chief engineer and assistant, 96; engineer of the steamer, 70, 71, 223; full-timers, 184, 185, 190, 277; part-timers Fort Howard, 236

Scheller, Herman, 110i

Scheller, Louis, 127, 261, 262, 264, 269

Schumaker, Mrs., 133

Shaler, Ephraim, 12-16, 208-210

Sheridan, Peter, 223-225,

Silberdorf, Conrad, 110i

Telephones, 109, 152, 174, 276

Triangle, alarm, 93, 212

Trumpet, speaking, 42i, 85i, 212, 296i-299i

Turner Hall, 189

US Army/military: quartermaster, 15; soldiers help at 1850 fire, 21, 22

US Hotel, destroyed, 62

Volunteer firefighters, 37, 177-181, 192, 193; benefits, 182, 208, 207i, 208; diploma, 292i; disbanded, 190, 190i; final inspection, 187-190, 187i; Germania resigns, 183; help at tannery fire, 199, 200; last Fort Howard part-timers paid, 279; numbers in Fort Howard, 235, 236; numbers in Green Bay, 98, 99, 173i, 180, 181,

Walters, John, 256i, 257i, 276i, 277

Washington House hotel fires, 25, 31

Water supply: access through ice, 48, 214, 234; bridge access, 202i, 214, 228i, 237i, 239; cisterns, 13, 66; cisterns in Fort Howard, 202, 215, 228i, 237, 236, 239; cisterns in Green Bay, 73, 74i, 79-82, 81i, 105, 113i; docks, 48, 71i, 80, 81i, 105, 113i, 214; failure during 1880 fire, 120, 121; rivers 69, 72, 73, 215, 214, 236

Index 349

Water Works: above ground reservoirs, 144i, 145, 150, 156, 164i; Britton fire Problems, 172-176, 172i, 175i; construction, 145- 150; formal tests, 150, 151i; Fort Howard pump station, 246; name, 144 279; notification, fire pressure, 151, 152, 276, 278, 279i; ordinance, 142, 144; pressure, normal/fire, 144, 145, 151, 245; supply pipes to Fort Howard, 144, 146-149, 162; system description, 144, 145, 146, 162, 240; South Adams facility, 144i, 145i, 147i, 163, 163i; Water Supply Committee, 162

Waupacaa Fire Department (Waupaca), 217

Weise, Albert, 40, 40i

West De Pere steamer, 131, 131i

Wheelock, Charles, 251, 253, 254

Whitney, Daniel, 5, 5i, 9, 14, 16, 18

Whitney School, 131

Whitney Square, 130

Wolff, Arnie, 315

Women, fighting fire, 63

Author Contact

Email David Siegel at:
gbfd.history@gmail.com